My Cat has Arthritis

– but lives life to the full!

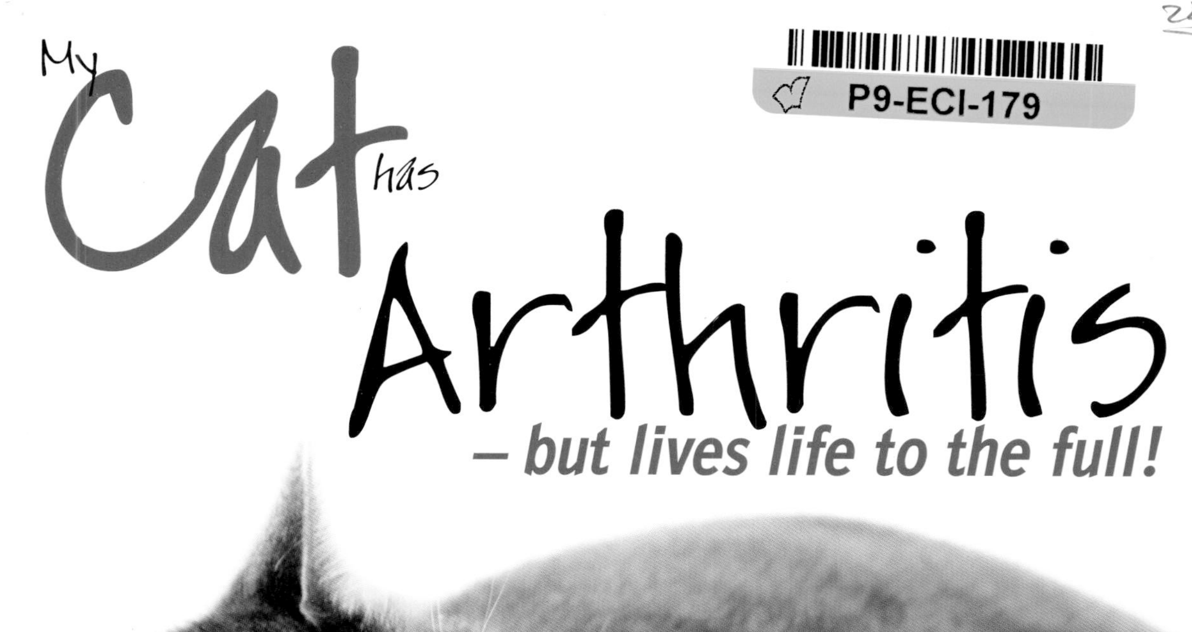

A practical guide for owners

Hubble & Hattie

The Hubble & Hattie imprint was launched in 2009 and is named in memory of two very special Westies owned by Veloce's proprietors.

Since the first book, many more have been added to the list, all with the same underlying objective: to be of real benefit to the species they cover, at the same time promoting compassion, understanding and respect between all animals (including human ones!)

Hubble & Hattie is the home of a range of books that cover all-things animal, produced to the same high quality of content and presentation as our motoring books, and offering the same great value for money.

More titles from Hubble and Hattie

A Dog's Dinner (Paton-Ayre)
Among the Wolves: Memoirs of a wolf handler (Shelbourne)
Animal Grief: How animals mourn (Alderton)
Camper vans, ex-pats & Spanish Hounds: from road trip to rescue – the strays of Spain (Coates & Morris)
Cat Speak: recognising & understanding behaviour (Rauth-Widmann)
Clever dog! Life lessons from the world's most successful animal (O'Meara)
Complete Dog Massage Manual, The – Gentle Dog Care (Robertson)
Dieting with my dog: one busy life, two full figures ... and unconditional love (Frezon)
Dinner with Rover: delicious, nutritious meals for you and your dog to share (Paton-Ayre)
Dog Cookies: healthy, allergen-free treat recipes for your dog (Schöps)
Dog-friendly Gardening: creating a safe haven for you and your dog (Bush)
Dog Games – stimulating play to entertain your dog and you (Blenski)
Dog Sense and CBT - Through your dog's eyes (Garratt)
Dog Speak: recognising & understanding behaviour (Blenski)
Dogs on Wheels: travelling with your canine companion (Mort)
Emergency First Aid for dogs: at home and away (Bucksch)
Exercising your puppy: a gentle & natural approach – Gentle Dog Care (Robertson & Pope)
Fun and Games for Cats (Seidl)
Know Your Dog – The guide to a beautiful relationship (Birmelin)
Life Skills for Puppies: laying the foundation for a loving, lasting relationship (Zulch & Mills)
Miaow! Cats really are nicer than people! (Moore)
My dog has arthritis – but lives life to the full! (Carrick)
My dog is blind – but lives life to the full! (Horsky)
My dog is deaf – but lives life to the full! (Willms)
My dog has epilepsy – but lives life to the full! (Carrick)
My dog has hip dysplasia – but lives life to the fuil! (Haüsler & Friedrich)
My dog has cruciate ligament injury – but lives life to the full! (Haüsler & Friedrich)
My Dog, My Friend: heart-warming tales of canine companionship from celebrities and other extraordinary people (Gordon)
No walks? No worries! Maintaniing wellbeing in dogs on restricted exercise (Ryan & Zulch)
Older Dog, Living with an – Gentle Dog Care (Alderton & Hall)
Partners – Everyday working dogs being heroes every day (Walton)
Smellorama – nose games for dogs (Theby)
Swim to recovery: canine hydrotherapy healing – Gentle Dog Care (Wong)
The Truth about Wolves and Dogs: dispelling the myths of dog training (Shelbourne)
Waggy Tails & Wheelchairs (Epp)
Walking the dog: motorway walks for drivers & dogs (Rees)
Winston ... the dog who changed my life (Klute)
You and Your Border Terrier – The Essential Guide (Alderton)
You and Your Cockapoo – The Essential Guide (Alderton)

Disclaimer

This book provides general information only, and is not intended in any way as a substitute for professional veterinary advice. Always consult your vet if you're concerned your cat might have arthritis. It is important to check with your vet before acting on any of the information given in this book as recommendations might change and new treatments could be introduced. All the information provided in this book was correct at the time of going to press: the author and Veloce Publishing Ltd shall bear neither liability nor responsibility with respect to any loss, damage or injury caused, or alleged to be caused directly or indirectly by the information contained within this book.

Photo credits

The photos have been supplied by individual owners, the author, or, where indicated, external sources. Illustrations by Natalie Knowles.

For post publication news, updates and amendments relating to this book please visit www.hubbleandhattie.com/extras/HH4618

www.hubbleandhattie.com

First published in June 2014 by Veloce Publishing Limited, Veloce House, Parkway Farm Business Park, Middle Farm Way, Poundbury, Dorchester, Dorset, DT1 3AR, England. Fax 01305 250479/e-mail info@hubbleandhattie.com/web www.hubbleandhattie.com
ISBN: 978-1-845846-18-3 UPC: 6-36847-04618-7
Readers with ideas for books about animals, or animal-related topics, are invited to write to the editorial director of Veloce Publishing at the above address. British Library Cataloguing in Publication Data – A catalogue record for this book is available from the British Library. Typesetting, design and page make-up all by Veloce Publishing Ltd on Apple Mac. Printed in India by Imprint Digital Ltd

Contents

Acknowledgements, Introduction and Dedication

Acknowledgements

Special thanks are due to Dr Richard Gowan of the Cat Clinic in Melbourne, for his expert input, patience and guidance on the subject of feline arthritis.

I'd also like to express my gratitude to Toby Gemmill at Willows Referral Services; Sarah Caney of Vet Professionals, and Jo Murrell at Bristol University for their help with case studies, and for kindly reviewing some of the chapters; to Sarah Clemson of Vet Physio and Brian Sharp of Canine Physio, and to Dr Pierson of catinfo.org for her input on the feline diet.

Thanks also go to Karen and Claire Bessant of International Cat Care (formerly Feline Advisory Bureau) for connecting me with feline experts, and for helping me generally; Samantha Lindley of the Chronic Pain Clinic, Glasgow University; Dr Jim Rummel and Jennifer at Camboro Vet Hospital, Pennsylvania; Natalie Finch, Bristol University; Graham Law, Glasgow University; Nick Thompson, the Holistic Vet; osteopath Tony Nevin of Zoo Ost Ltd; Claire Feek of Norwich RSPCA; animal behaviourist Steve Dale of the Winn Feline Foundation, and Steve Crow of the Governing Council of the Cat Fancy.

I'm also grateful to Helen Reeve of Avonvale; Phil and Debbie at Blossoms, and Sue Hawkins of Hawksmoor hydrotherapy centres; the PDSA press office; Ken at Norfolk Pet Crematorium Ltd; the Celia Hammond Animal Trust; Wood Green Animal Shelters; Gay Rees of the North Norfolk Cats Lifeline Trust; Bengal expert Barrie Alger-Street; Sue Parslow of *Your Cat* magazine, and Alex Hurrell of Eastern Daily Press (EDP) newspapers.

Finally, I owe a huge debt of gratitude to the marvellous owners who have shared their cat's inspiring stories with me, and sent through some lovely photos of their firmest feline friends enjoying life to the max.

Introduction

Just like people and dogs, cats are living longer, with many of our feline friends chalking up, on average, around 12 years, and a few reaching the impressive numerical heights of 20 and beyond. And while living to a ripe old age is generally a good thing for cats (and nice for their owners), the downside is that the older they get, the more likely they are to develop osteoarthritis, or degenerative joint disease.

Arthritis is a common problem in cats, with most felines over eight suffering from the disease to some extent. It's a painful condition – one of the main causes of chronic pain in cats – and can badly affect their quality of life and ability to enjoy the fun activities they so love. It can have a primary root – as a result of wear and tear over time – or a secondary cause, perhaps as the result of an injury, or a joint defect.

As cats are rather good at hiding any pain or discomfort they're experiencing, it can be difficult to tell if your cat's suffering – but the signs are usually there if you know what to watch out for. If he's reluctant to jump onto the sofa, for example, there's a good chance painful arthritis is to blame.

Of course, if cats could talk they could tell us that their joints hurt, and ask for help, but, until then, *you* have to be your cat's advocate, ensuring she gets the help she needs and deserves. Your cat will thank you for it, from the tips of his paws to the tops of her ears.

Hopefully, this book – with expert veterinary advice; heart-warming case histories; chapters on treatment options; tips for coping, and recommendations about diet and supplements – will help you to help your cat. There are also sections on physiotherapy; complementary therapies; hydrotherapy, and surgical options if your cat's arthritis warrants it, plus useful contacts.

It's always worth sharing your stories with other owners by joining a local club, or by taking part in one of the increasing number of online forums for specific breeds, which discuss everything from concerns over health to the latest pet aids and health supplements. New social networking sites, purely for cat owners, are springing up, too. Some nice vets post comments and advice online, so you could well pick up an extra tip or two, free of charge, which can't be bad.

Finally, stay positive: chances are your cat will, as our case histories show. Here are just a few of the furry felines we'll be meeting along the way –

Dedication

To all the cats who haven't let a little thing like arthritis stop them in their tracks; including my own furry friend, Sweepie, who enjoyed his golden years in France.

Foreword

Cats have become increasingly popular household companions, and in many countries cat ownership now outweighs that of dogs. However, it is still very much the case that cats are not taken to a vet for regular and investigative health checks as often as dogs. This is for many reasons, including the perception that cats are hardier and do not require the same level of care – and the notion that cats mask the signs of illness and painful disease, such as osteoarthritis, merely as preservation behaviour.

However, over the past fifteen years I have spent in feline-only practice, a tremendous amount of research and focus has led to the realisation that cats, especially with increasing age, indeed suffer the effects of painful, chronic, musculoskeletal disease. And not only are the majority of aged cats suffering from this degenerative condition, it is a truly painful and debilitating one. Yet, with thorough health investigation and multi-modal pain management strategies, the disease progression can be slowed, the pain minimized, and the quality of life of cats around the world improved.

I have developed a professional interest in feline arthritis and chronic pain management. Our practice has published several articles in the *Journal of Feline Medicine & Surgery*, including a review of the management outcomes for our patients with arthritis, and the longest study into the long-term results of the use of pain relief for cats with the disease, and how it has affected their longevity. I have had the great pleasure of lecturing many colleagues around Australia and in Europe, sharing my passion for cats, and the need to challenge long-held beliefs about feline arthritis, pain, and how it impacts our feline patients. Of recent focus has been a review of how chronic pain in all disease states influences cat behaviour.

Many a cat owner would report their cat as 'slowing down' or 'getting old,' but would never make the connection that these changes in behaviour were as a result of, or to compensate for, arthritic pain. Cats live how they feel, shaped by their individual needs, wants and desires. Pain is an inhibitory experience that frames their motivation: namely to weigh up the consequences of pain versus the desire for pleasurable experiences. Old, normal behaviours change gradually and eventually stop, and new, abnormal behaviours are developed – so reducing discomfort.

Raising awareness among cat owners about how cats adapt their behaviour to reduce their pain – coping mechanisms we thought previously were to help to mask their illness and pain – is the key to combating this 'silent suffering' of cats.

This book shares the insights and practical management perspectives of cat owners and health care professionals about this complex disease. Books such as this are an invaluable resource for cat lovers, and undoubtedly will lead to the improved quality of life – and potentially the longevity – of countless feline friends across the world.

Dr Richard Gowan
BVSc(Hons) MACVSc (Feline Medicine)
The Cat Clinic, Melbourne, Australia
President of the Australasian Society of Feline Medicine

Chapter 1

The different types and signs to look for

Big cats living in the wild and domestic cats sharing our homes have the same survival instinct. Adept at masking any pain or discomfort they might be suffering, they find somewhere private to lick their wounds: out of sight of any potential 'predator.' This preservation behaviour makes it difficult to diagnose arthritis in our feline friends.

While the signs of arthritis are easier to spot in dogs – as they're around us more – cats tend to spend a lot of time alone, making it harder for owners to detect any problems – or, indeed, to believe that any exist.

But arthritis is definitely not the sole preserve of dogs (or humans): a recent study which looked at the x-rays of a hundred cats, found that 90 per cent of them had radiographic signs of arthritis. So arthritis is a more common problem among cats than we probably realise.

Cats of any age can develop the condition, but the older (and more overweight) your cat is, the more likely he or she is to develop the disease. With so many cats now living well into their teens and beyond, it's a problem that's surely set to grow.

What is feline arthritis?

Arthritis is a catch-all term for a group of conditions affecting not just the joints (which stiffen and become inflamed), but also the muscles, tendons, cartilage, bones and ligaments supporting the joints. It can develop in a cat's shoulder, hips, or hind legs – where 60 per cent of a cat's body weight is carried – and in the elbows, knees (stifles), and ankles (tarsi); it's part of a condition known as degenerative joint disease, or DJD.

Degenerative joint disease (DJD)

DJD is a complicated condition in which the cartilage that cushions the joint degenerates, causing pain, damage, and secondary changes in and around the joint. Cartilage is one of the key tissues in a healthy joint, acting as a shock absorber and providing a smooth surface between the bones. It's very slippery – twenty times more so than ice – and consists of collagen fibres filled with hyaluronic acid and chondroitin sulphate, among other components.

What causes DJD?

Other than wear and tear over time, for

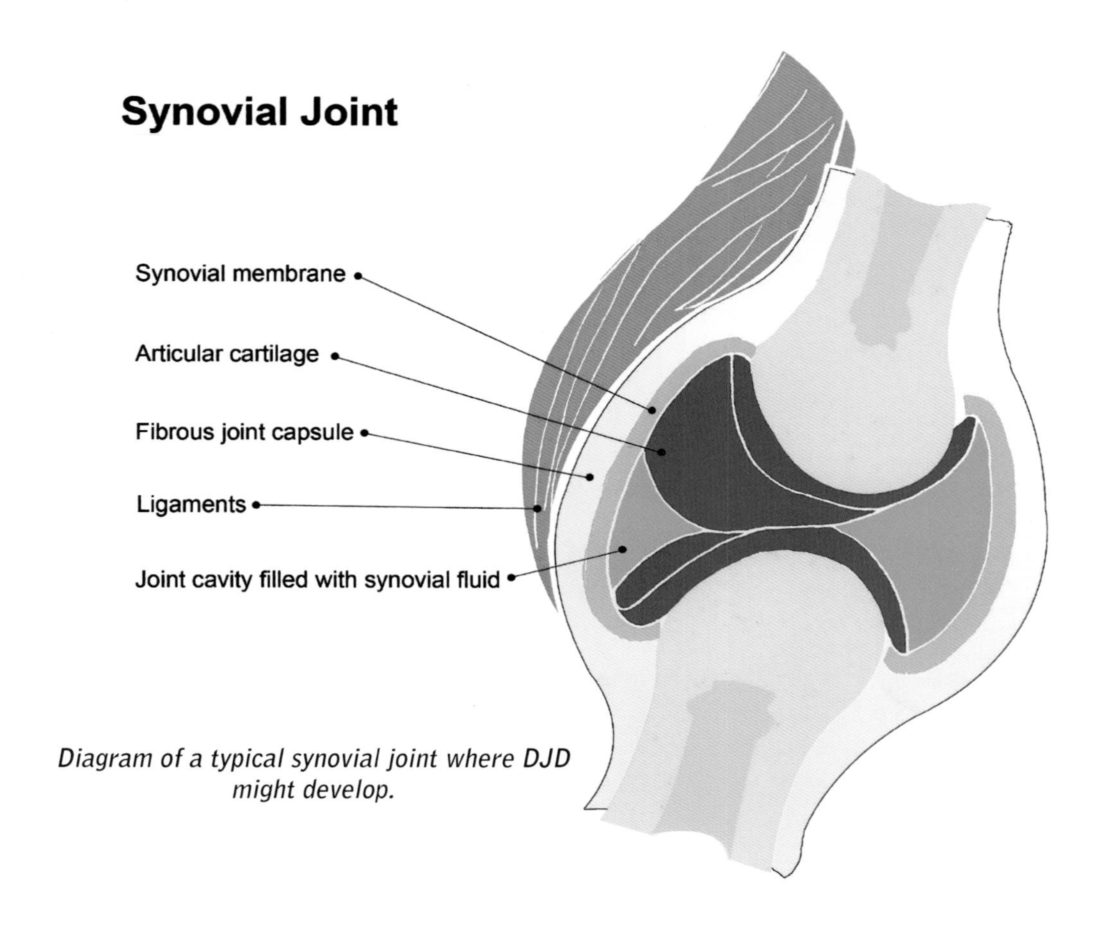

Diagram of a typical synovial joint where DJD might develop.

reasons which aren't always clear, changes occur in the structure of a joint, causing inflammation, which sets off the disease process.

As the affected joint becomes stiff and swollen, the cartilage is overloaded, and can no longer do its job of cushioning the ends of the bones, which then grate against each other. Moving around becomes painful and difficult for your cat as a result.

And while being overweight doesn't cause DJD, it certainly puts extra pressure on your cat's joints – especially over time – making the problem worse. Genes play a part, too. The main kind of degenerative joint disease in cats is osteoarthritis, or OA.

Spotting the signs of OA

It can be hard for an owner to spot the clues as they can be non-specific, and even overlap with a number of other diseases that can be common in older cats. And unlike dogs, cats don't usually limp – one of the tell-tale signs of osteoarthritis in dogs, for example – or hold up their painful paw for maximum sympathy value. Which means that, even as the animal's carer, you might not have an awful lot to go on. Added to which, cats can cope well with arthritis, with running and horizontal movement not changing a great deal in cats with OA.

Pointing the way

However, any change to how high your cat jumps – as well as his willingness to jump and the frequency of doing so – are the most specific indicators of the pain of OA. So, if your cat shows reluctance, or refuses, to jump onto the bed or sofa as he would normally do, for example, or if he finds it difficult to go up or down stairs, painful osteoarthritis could be at the root of the problem.

What else could suggest OA?

- Struggling to get in and out of the cat flap
- Spending more time resting or sleeping
- Not hunting or exploring outdoors as frequently
- Sleeping in different and easier to access places
- Being less interested in playing
- Spending more time alone

Changes in grooming habits

Cats suffering arthritic pain might spend less time grooming, and groom less often, or not at all, which could leave their coats matted. (They might also veer in the other direction and over-groom painful joints.) A cat with OA might spend less time sharpening her claws, so these are likely to be longer than they should be.

Changes in grooming patterns can also be caused by other problems, or because of a cat's advancing age, so is not a definite sign of OA.

Changes in behaviour

When cats are in pain, their behaviour is likely to change, and they might not do certain things that they've always previously done because it's too painful to do so. For example, your cat might begin to have accidents around the house, avoiding the litter tray, perhaps, and using the more convenient, and possibly softer, carpet instead. Or, when trying to stroke your cat she might make it plain that she doesn't welcome the attention, and may even be aggressive toward you.

Changes in normal behaviour, and the

Sweepie liked to be lifted onto the sofa, rather than having to jump up.

Mouseski sleeping off playtime.

A cat showing signs of distress could be in pain.

development of new, abnormal behaviour, may be an indication that your cat has an underlying disease, which could be one of a number of conditions – including painful OA.

Other types of arthritis

Although very rare – your vet might see just one case in every ten years – a cat can suffer from what's known as 'immune-mediated' arthritis, whereby his immune system malfunctions, going on the attack. (This kind of arthritis is similar to rheumatoid arthritis, or lupus, in humans.) In this instance, a cat's entire system – including his eyes and kidneys – is affected, as well as a number of joints (polyarthritis), rather than just one. It's usually a condition for life and could be due to an underlying problem such as a tumour, or a virus.

Getting a proper diagnosis

As it's hard to spot the definite signs of osteoarthritis, you might have to rely on your instinct that something is wrong with your cat, and needs investigating. It's a good idea to pay a visit to your vet as soon as possible to get a firm diagnosis of osteoarthritis, and rule out any underlying problems unrelated to OA.

Managing arthritis

While it is inevitably distressing to learn that your cat has arthritis, the good news is that a lot can be done to relieve the pain and discomfort your cat will be experiencing.

As explored further in separate chapters, the best way to tackle the disease is with a combination of good veterinary care; conventional drugs to relieve the pain and help control the inflammation; dieting your cat if she's carrying around too many extra pounds, and making any necessary changes at home – particularly if your cat is getting a bit long in the tooth.

Physiotherapy will help to maintain mobility, particularly after injury, and some complementary therapies such as acupuncture and osteopathy could help to ease the symptoms of OA. In no time, the spring will be back in your cat's step, and her tail will be up in the air.

Case history: Pebbles

Pebbles, 13, was diagnosed with OA a couple of years ago, says her owner Amanda –

"We noticed Pebbles was hiding under the bed a lot, which is what she normally does if she's unhappy or ill, and she seemed reluctant to jump up. We thought at first that maybe she'd been injured by a seagull, as they were dive-bombing the garden because one of their chicks was there, but we weren't sure and decided the best thing to do would be to take her to the vet and get her checked over. An x-ray showed her hips had the first stages of osteoarthritis.

"We manage her condition with meloxicam and seraquin, which I sometimes increase if she's looking a bit creaky. The vet suggested hydrotherapy, but she hasn't tried that yet.

"Pebbles still runs around the house like a mad thing, and she adores playing with her toys. She loves the heat, and as soon as it's winter and the radiator in the lounge is turned on, she snuggles up against it. My late cat, Bows, who also had arthritis, lived to nearly 18, thanks to medication and diet, and nothing is going to stand in Pebbles' way."

Pebbles is still living life to the full.

Chapter 2

Joints affected

Why an ever-increasing number of cats are developing osteoarthritis (OA) isn't always clear, but, as a general rule, a mixture of nature (genes) and nurture (a trigger such as an injury) lies at the root of the problem. And the older a cat gets, and the more weight he's carrying around, the more likely he is to be affected by the condition at some stage in his life.

As in dogs, osteoarthritis in cats generally occurs as a secondary problem to defects in a joint, such as hip dysplasia, a fracture in a growth plate, or a ligament tear. OA can also come about as the result of osteochondritis dissecans (OCD), which affects the cartilage in some joints. Some older cats suffer from primary OA which occurs as a result of wear and tear in the absence of an underlying injury.

Hip dysplasia (HD)

Although recognised as a common problem among many dog breeds, cats were thought not to suffer from HD, yet research has found that HD does indeed exist in cats; particularly among the larger, heavier-boned breeds, although smaller breeds and mixed-breed cats can be affected, too.

What is hip dysplasia?

A cat's hips consist of two ball and socket joints – one for each leg – and, rather like a caravan tow bar, the joints should fit snugly, allowing a cat to walk normally. In cases of hip dysplasia the joint is often too shallow, and the ball too small and out of shape to fit tightly, so the muscles, ligaments and tendons around the joints become loose, with the result that the hip joint dislocates and becomes unstable.

HD is not necessarily obvious at birth, and normally occurs during a young cat's growing stages. Often, if one hip is affected, the other one will be, too. Over time, chronic changes develop in the bones of the hip joint due to this abnormal movement, and osteoarthritis might result unless the hip (or hips), are replaced. Dysplasia can also develop in the feline elbow.

Possible causes

Recent research has shown that HD is probably an inherited disorder in cats, as it is in dogs. If a cat's found to have hip dysplasia, then one or both parents must either be affected, or be carriers of the defect.

HD is more common in a number of larger breeds than at first thought.

Aside from a genetic connection, if a kitten has grown rapidly, or perhaps hasn't been fed a balanced diet from the start, the likelihood of HD developing increases. An injury, however slight, to a cat's hip at important stages of his development is another risk factor.

Symptoms

These vary according to a cat's age, and range from appearing a little clumsy as a kitten to having difficulty running or climbing stairs from around one year.

Diagnosis & treatment

HD is usually diagnosed with x-rays and sometimes CT scans, and treatment might centre on painkillers and non-steroidal anti-inflammatories (NSAIDs), although, in most cases, surgery to replace the damaged hip and restore full function in the joint is the only course of action that will spare a cat from suffering a lifetime of pain.

HD can start to develop in kittenhood.

Other hip problems

A cat's hip is subject to more problems than any other joint in her body, and, unfortunately, other conditions can occur, aside from hip dysplasia.

A relatively common problem in young cats, particularly neutered males, is a condition known as slipped femoral capital epiphysis, or SFCE: a fracture which develops over several weeks through a layer of cartilage, or growth plate, at the head of the femur (thigh bone) within the hip joint. (A weakness in the growth plate is believed to be at the root of the problem.)

Treatment options for SFCE are limited, and, as the fracture can't be effectively stabilised, affected cats develop severe arthritis of the hip. Therefore, surgeons will either remove the head and neck of the thigh bone, or perform a total hip replacement, as in Alfie's case (page 18).

Cranial cruciate ligament (CCL) injuries

The knee joint of a cat is one of the weakest in her body. There are no interlocking bones in the joint, so it's held together by several ligaments, including the cruciate (crossover) ligaments that connect the femur (the upper bone of the leg) to the tibia (the lower leg bone).

The ligaments allow the knee to flex backward and forward, but restrict side-to-side motion. When a ligament tears, causing instability, the bones then move in an irregular way to each other, and it's difficult for a cat to bear weight on the leg without it collapsing.

Possible causes

A tear to a CCL can occur when a cat experiences a trauma, such as being hit by a car; when he jumps or runs, or if he falls from a significant height, perhaps from a high-rise flat (often referred to as 'apartment syndrome' in the US). If there have been degenerative changes to the collagen fibres around the joint as a cat gets older or gains weight, this can increase the risk of CCL tears.

Symptoms & diagnosis

Signs of a CCL tear can include the rapid onset of rear leg lameness, which will need to be confirmed by your vet, who will examine the knee to see how stable it is (your cat might need to be sedated for this), and then take x-rays.

Treatment

Although painkillers and rest usually help with the problem initially, most vets will suggest surgery if your cat does not improve, in order to stabilise the knee and reduce the severity of osteoarthritis further down the line.

Repairing the damaged knee might help to prevent similar tears in the cruciate ligaments in the other knee. Helping your cat to lose any extra weight will speed recovery time following surgery.

Osteochondritis dissecans (OCD)

OCD – which is also referred to as osteochondrosis – is a rare condition in cats whereby the cartilage around the bone in a joint doesn't form properly. The result is a loose piece of cartilage that flaps around, restricting movement and causing pain, and allowing osteoarthritis to develop in the affected joint over time.

A number of joints can be affected by OCD, including a cat's elbow, shoulder, knee and ankle. If the loose cartilage breaks free it will float around in the joint (described as a 'joint mouse').

Symptoms of OCD

If your cat appears lame, particularly after exercise, this could be a sign of OCD. Your vet will need to carry out a full examination

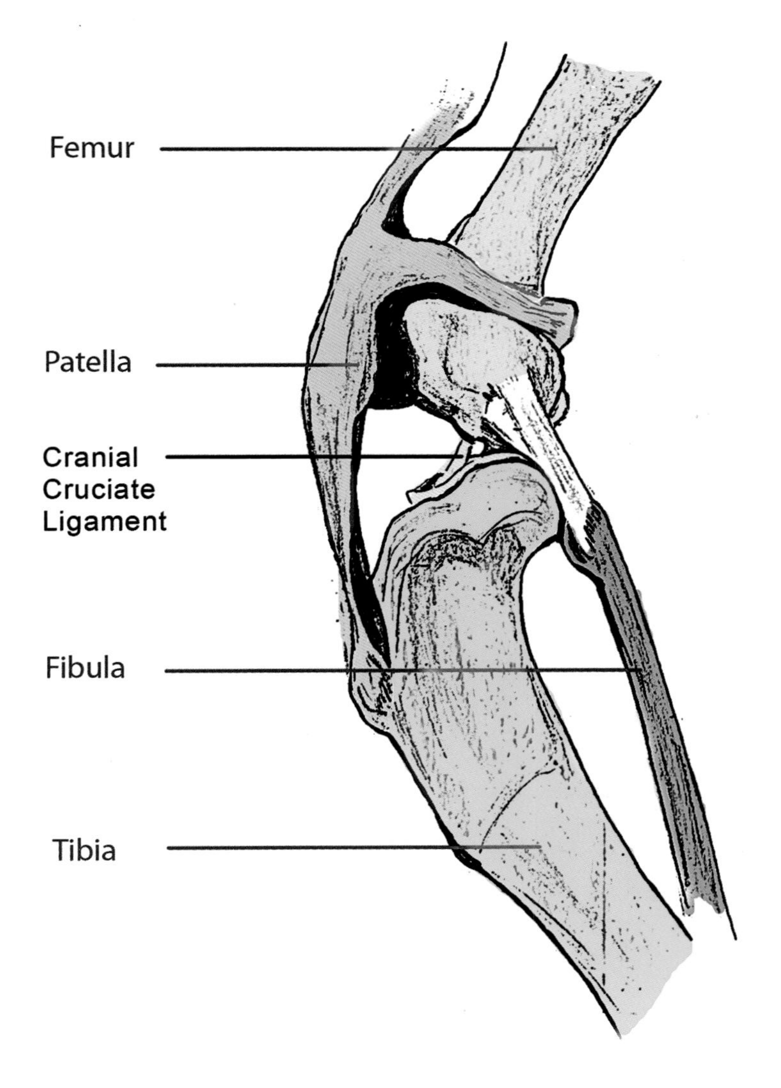

A cranial cruciate ligament can easily tear, leading to OA later on.

and take x-rays, and perhaps an MRI or CT scan, to confirm the diagnosis.

What causes OCD?

The exact cause is unknown. Whilst OCD has been linked to certain breeds in dogs, this doesn't seem to be the case with cats, who might simply be born with defective genes, leading to abnormal development in joint cartilage. For some, it might be the result of some kind of trauma, or deficiencies in their diet.

Treatment

Your cat might be prescribed non-steroidal anti-inflammatories (NSAIDS) and reduced exercise at first, which might be enough to reverse the problem. But if the lameness hasn't gone away six months or so of treatment, surgery will probably be needed to remove the loose flap of cartilage, as well as reduce the likelihood of osteoarthritis developing later on.

Alfie's back to his old self at home.

Case history: Alfie

Alfie, a domestic short-hair male, showed signs of lameness in one leg initially, but by the time he was two years old, both legs were clearly affected. His vet decided to refer him to Willows Referral Services in the West Midlands for specialist treatment.

An examination and x-rays showed Alfie had fractures in both his growth plates, and he was clearly in pain. The surgical team at Willows decided to replace both his hips, using cemented implants. His right hip was replaced in 2010; his left two years later.

Alfie made an excellent recovery from both operations, and is now doing very well: he's able to walk, run and jump without pain. Without this surgery, Alfie would have been in chronic pain. His owner, Biba, is delighted with the outcome –

"I think, in some ways, it was better Alfie had the operations while he was young so he could recover easily. He had to be in his cage for a while after each operation, and had to be supervised for a good few months, but he never complained. He's five now and you wouldn't know he'd had anything done.

"He's an indoor cat and loves jumping, running and trying to catch bouncing balls – his favourite thing to do. He was quite a celebrity for a while at Willows, and a bit of a pioneer, as he was the first cat to have a full hip replacement at the veterinary practice. I've had Alfie since he was a kitten, and he's my pride and joy: a very loved little boy, who's now doing so well."

With thanks to surgeon Toby Gemmill, a specialist in small animal surgery at Willows Referral Services.

Chapter 3

A diagnosis – what now?

If you suspect your cat has osteoarthritis, the best thing to do is to pay a visit to your vet for a definite diagnosis. The earlier the confirmation of the disease (if this is indeed the problem), the sooner any treatment can begin, and your cat can resume enjoying her life to the full.

A thorough check

Typically, a vet will conduct a full physical examination, looking for any swelling in your cat's joints; thickening around her knees and elbows, and any obvious signs of stiffness and pain. How your cat's standing – perhaps putting more weight on a particular leg, say – will also be noted.

X-rays might be ordered to ascertain whether there's any joint damage, or hip problems, although these might not show up in the early stages. Many vets will recommend blood and urine tests before starting treatment with a non-steroidal anti-inflammatory such as meloxicam, since cats with renal (kidney) disease may be more susceptible to side effects.

Your role

As the closest person to your cat, you'll probably be asked about her daily routine and how it may have changed. For example, is your cat jumping up as much as she used to? Where does your cat sleep – on the sofa, on your bed, by the radiator – and for how long each day?

You might also be asked if you have more than one cat at home, and whether they play together, and if your cat's mood or eating habits have altered recently.

Tips for coping

If osteoarthritis is confirmed, this isn't as bad as it might seem, as cats generally cope well with the condition.

You could try giving her a gentle massage to increase flexibility and circulation: most cats, with careful handling, respond well to this.

You could also consider physiotherapy, hydrotherapy and supplements, which might help manage some of the symptoms, alongside conventional treatment (see separate chapters).

And there are a number of practical measures you can take at home to help maintain your cat's quality of life.

A good bed

Every arthritic cat deserves a bed that's

Bamboo seems happy in her bed. A bed with a removable cover makes keeping it clean much easier. (Courtesy 26barsandaband.com)

comfortable, easily accessible, away from draughts, and with padding that's not so deep he has to struggle to get out. Some owners opt for 'igloo' beds, which can make an older cat in particular feel warm and secure.

Or you could consider investing in an orthopaedic bed with special pads to distribute your cat's weight evenly, particularly if he's a little on the heavy side.

A Flectabed can help to keep little joints warm. (Courtesy Petlife)

Cat flap

Make sure the cat flap is easy to open, and if there aren't too many other cats in the local area who might be tempted to pay a visit, it might be worth tying it open so that your cat doesn't need to push through it.

Litter trays

A small litter tray might be a tight fit for your cat, so always chose a litter box which is large and easy to get in and out of: a flatter, pan-style box might be a good idea. And make sure it's in the right place: if your cat lives mainly on the top floor, don't put her litter tray in the basement, or she might not bother using it – and who could blame her?

Cosy corners

Most cats seek out the warmest part of

the house, which is often in front of the fire, or by a radiator, so you could invest in a special carrier that attaches to most radiators. Some owners find wrapping a hot water bottle in a fleece for their cat to curl up on works a treat, and in no time at all your moggie could be in cat heaven, dreaming of mice, or anything else he's had his eye on in the garden!

Heated pads

Heated pads can provide up to ten hours of warmth after being microwaved for a few minutes, and can be placed anywhere. They're often used by vets in post-op

Mother and daughter, Annie and Tila, loved cuddling up together in the warmest spot in the house.

A Snugglesafe heatpad. (Courtesy Lenric C21 Ltd)

recovery and convalescence, and might help to make life a little more comfortable for your cat while warming her joints nicely.

Ramps

If your cat's finding it difficult to jump onto the sofa, the windowsill or your bed, why not consider a ramp or steps to help him? They're usually lightweight, are often foldable, and will reduce any impact on his painful joints.

Elevated feeders and water bowls

Bending to eat and drink is not comfortable for cats with muscle and joint problems. Older cats often don't eat as much as they used to, anyway, and if eating and drinking are painful or uncomfortable, they're likely to eat or drink even less, which isn't good for them. So it's worth considering raising your cat's food and water bowls for easy access, so he doesn't have to position his head downward.

Clinics

A number of veterinary practices offer arthritis clinics, which are often run by one of the veterinary nurses. They're usually free (although any treatment will have to be paid for), and can help before, and after, a diagnosis.

The clinics offer owners the chance to chat about how their cat is coping generally; whether the drugs are working; if the amount can be reduced and when (although older cats in pain can't come off as quickly or easily), and if there are any changes, or increase, in symptoms.

Grooming

If your cat's finding it hard to groom certain areas, perhaps because her spine or hips are hurting, her fur could become knotted or matted.

You could try to groom these areas

Sherri being professionally groomed by Nicola at Mr Schnorby's in Cromer.

for her, or, alternatively, take her to a qualified groomer. Some cats tolerate this reasonably well, although most need to be sedated by a vet first; particularly if they're easily stressed, or tend to bite or scratch when in strange surroundings.

Cat-friendly clinics

The cat-friendly clinic initiative, a scheme run by the International Society of Feline

Case history: Mouseski

Rachel, Mouseski's owner says –

"Mouseski was diagnosed with severe osteoarthritis, and, in theory, shouldn't have been able to walk, but he was so agile he climbed six-foot walls, and chased off other toms in the street! He went on to develop a large, fluid-filled lump on his left shoulder, but this didn't stop him from doing anything, and he amazed all the vets.

"He was on meloxicam, and the vet showed me how to massage his leg to help his circulation and relieve any pressure, as the lump was causing fluid to build up in his leg. I did this several times each week, which helped him a lot. I also made several changes to his environment that I noted by monitoring him daily.

"Mouseski loved his raised beds and chairs, and because he loved them so much, I provided memory foam landing spots underneath the chair and the raised beds so that when he jumped down, his landing was cushioned and his joints were protected from any strain or pressure.

"Mouseski also loved his food, and, because I knew that he must be in a lot of pain and in great discomfort, I elevated all of the bowls, which certainly did the trick as he was able to drink water and eat his food all around the house without putting any pressure on his two front joints, which were the worst affected. The transformation was incredible and you could see how content he was when he ate and drank.

"I decided to put the garden chairs next to the lowest part of the wall in our garden to allow him to jump on to the chair and then to the floor, instead of jumping straight to the floor. I used treats to teach him where to jump from, so he knew it was better to jump at the lowest part of the wall, instead of the highest part. Mouseski was brilliant at this.

"Sadly, he's no longer with us as he had to be put down because of bowel cancer in 2012. He was a little fighter who stole my heart from the very first day I saw him."

Medicine (ISFM) and Purina pet foods, aims to improve the experience of cats at veterinary practices in the UK, and across the world. The initiative encourages practices to set up a separate cat-only waiting room or area, or cat-only consulting times and cat-only wards. (Visit www.icatcare.org to see a list of accredited veterinary practices in your area.)

Eating was easier for Mouseski when his bowl was elevated.

Case history: Ebony

Dr Sarah Caney of Vet Professionals runs a clinic for geriatric felines, which is based at the Royal (Dick) School of Veterinary Studies in Edinburgh. One of her patients, Ebony, is a typical elderly cat with osteoarthritis. Says Sarah –

"I saw Ebony as an 18-year-old domestic short-hair who was presented for assessment and management of her hyperthyroidism (overactive thyroid gland). When she was walking around the consulting room, it was obvious that Ebony also had mobility problems, as she walked with her elbows abducted (held away from her body, as shown in the accompanying picture), and had a very stiff and stilted, short-strided gait.

"At home she was reported to be reluctant to jump onto the sofa or bed, and was very slow going up and down stairs. When I examined her elbows I found that their range of movement was reduced, and she was unhappy for me to manipulate them. X-rays confirmed osteoarthritis, which was especially severe in both elbows.

"Successful management of Ebony's arthritis involved a combination of treatments aimed at improving her quality of life. Changes to the home environment included raising food and water bowls to make eating and drinking more comfortable; offering a low-sided litter tray that was easy to get in and out of, and comfortable beds at ground level. One such bed was a pillow covered with a fleecy blanket, located next to a radiator. We also prescribed some joint supplements containing glucosamine and chondroitin (although, anecdotally, these can be helpful in some cases, there is no conclusive proof of their benefit in cats). Lastly, Ebony received a low dose of painkiller every day to help keep her comfortable."

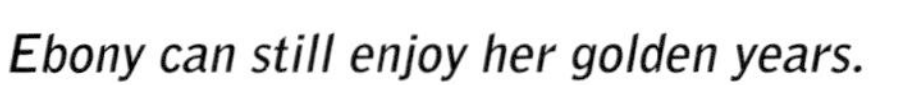

Ebony can still enjoy her golden years.

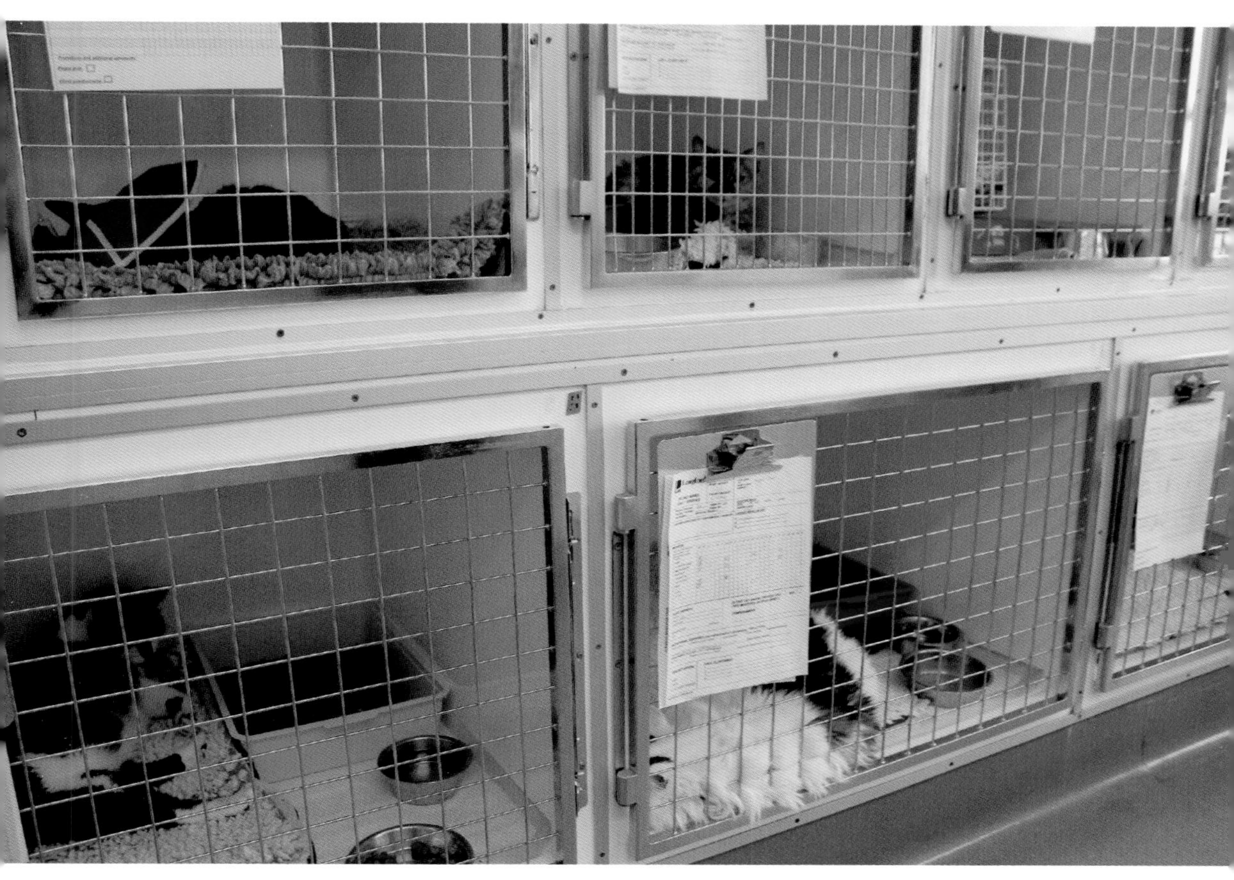

A cat-only ward at Langford Veterinary Services Small Animal Hospital, University of Bristol. (Courtesy Teresa Labiak)

Chapter 4
Treatment

While there's no magic bullet for arthritis, there are treatments available that will manage the inflammation and pain your cat will be experiencing, and improve quality of life. And with the right drugs; weight loss if needed; complementary therapies and adjustments at home to make your cat more comfortable, hopefully, in no time at all she should be back to living her life to the max.

Prescribed drugs

It wasn't so very long ago that cats were regarded as little more than small dogs, and scant attention was paid to feline-specific pain relief. Happily, that situation has now changed, and there are a number of drugs your vet can prescribe to help your cat, including –

- Analgesics (painkillers) such as non-steroidal anti-inflammatory drugs (NSAIDs), buprenorphine, tramadol, gabapentin or amantidine
- NSAIDs such as meloxicam (Metacam®), are not usually prescribed for cats with kidney or liver problems, but are very effective in managing pain and inflammation

Inflammatory arthritis

Cats with rare inflammatory arthritis, driven by a faulty immune system, will need treatment with powerful drugs for longer, as these kill the cells that are dividing quickly.

Treatment usually starts with a high dose, which is then reduced to the lowest possible level to manage the condition. Sometimes there's complete remission, but your cat might still require regular low level treatment (perhaps a dose every other day) to control the symptoms.

Side effects

As with all drugs for whatever type of arthritis, there can be unwanted side effects, and some cats – particularly those with certain diseases, or older cats – are at increased risk of adverse effects. It's understandable you might not be too happy about your cat having to take powerful drugs, but there's usually no effective alternative, particularly in the early stages of treatment and in severe cases, and side effects aren't necessarily automatic.

Once the inflammation is under control, the level of medication can be reduced gradually over time (although some cats don't cope too well if medication is reduced, or stopped altogether).

Always raise any concerns you might have with your vet, and discuss whether there's an alternative drug which might suit your cat better – after all, every cat is different – with fewer, or more tolerable, side effects.

Ask which would be easier to administer – tablets or medicine – and don't be afraid to ask your vet for advice on the best way to give the drugs to your cat – maybe even requesting a demonstration?

Ideally, cats should take their medicine with meals.

Meal times

Always give your cat the drugs with, or after, food, as defined by your veterinary surgeon. Some drugs taste bitter to cats, so disguising them in his favourite food might help. (My cat was way too canny for this ploy, and would usually just eat the food and leave the medicine well alone.) If your cat refuses to take his medicine at home, however hard you try and with whatever cunning tactics you employ, then it's worth asking your vet to suggest an alternative.

Importance of monitoring

Your vet might recommend, and perform, regular kidney and liver tests to see how your cat's body is tolerating a certain drug, and might make adjustments to find the lowest effective dose, or switch to an alternative therapy if necessary.

You have a role to play in being vigilant, and watching for possible side effects that your cat might be experiencing, which could include –

- Loss of appetite: a cat shouldn't go for more than one to two days without eating
- Vomiting
- Diarrhoea
- Lethargy

If your cat seems apathetic and lacking in energy, mention this to your vet in case her medication is responsible.

- Increased thirst

If your cat is suffering from any of the foregoing, tell your vet as soon as you can, and stop administering the treatment (veterinary recommendation).

Drugs meant for humans and dogs

Cats metabolise painkillers and NSAIDs differently from humans, so please don't be tempted to give your cat something like Ibuprofen, for example, which can be toxic for cats, or anything else that might be lying around in your bathroom cabinet. Only give your cat medicine that's designed for cats: products meant for canine use can be very dangerous to cats.

Controlling costs

Prescription drugs can be expensive, and it's usually cheaper to buy the drugs online rather than from your veterinary practice. Make sure you use a reputable online site, though, which insists on a valid prescription. Your vet will be able to supply one, although there's likely be a charge for this.

It's also worth investing in a good pet insurance plan to cover most (or all) of the costs should your cat need surgery, expensive treatments or tests.

Visits to the vet

If your cat has osteoarthritis, it's likely you'll need to see your vet on average around 3-4 times a year after diagnosis (or more often if symptoms worsen). Cats with inflammatory arthritis are more difficult to treat, and will need more veterinary attention – possibly for life.

In cases where your cat's arthritis is outside your vet's realm of expertise, or your cat is not responding well to the current treatment plan, your vet might refer your cat to one of the excellent veterinary referral practices up and down the country, which can handle more complicated cases and perform surgery, such as hip replacements, if necessary.

Referral to a specialist chronic pain clinic is also an option when pain management is challenging.

When surgery's the only option

If your cat's joint is severely damaged, or if the pain is intense, surgery might be the only way to go. While this is not without risk – and isn't usually suitable for overweight cats or those with underlying health problems – surgery can reduce pain and improve movement, and is the nearest thing to a cure in some cases. There are different kinds of surgery, ranging from a total hip replacement to more minor procedures, such as –

Arthroscopy

This is the least invasive type of surgery, involving the surgeon making small cuts over the joint to clean out cartilage damage in shoulders, elbows, and knee joints.

Joint replacements

Replacing a damaged joint is an option for an otherwise healthy cat who is suffering from painful hips and lameness. In the case of hip dysplasia for example (see chapter 2), the prosthetic (artificial) cup and ball joint is usually held in the pelvis and thigh bone by bone cement. An artificial hip will normally outlive your cat.

Stem cell therapy (SCT)

Adipose stem cell therapy introduces new cells into tissue damaged by osteoarthritis, and has been used on horses and dogs, though rarely on cats. While not without controversy and its critics, SCT has proved of benefit in some cases.

Case history: Luca

Luca's osteoarthritis is being treated at Langford Veterinary Services' Small Animal Practice, part of the University of Bristol, as vet Jo Murrell explains –

"Luca's owner, a veterinary nurse at the practice, noticed he was lame on the left leg, and less willing to jump up onto the sofa and bed; he was also over-grooming the inside of his leg.

"Luca had sustained a fracture to his lower tibia (shinbone) after a traffic accident when he was young (he's now ten), and this had been repaired with a pin (placed in the middle of the fractured bone), and two wires placed around the bone to stabilize bone fragments. Luca's osteoarthritis (OA) was confirmed by X-rays, which also indicated that the wires placed to stabilize the bone fragments might now be causing irritation and inflammation to the joint.

"Luca was given pain relief in the form of the non-steroidal anti-inflammatory (NSAID) meloxicam, and surgery was planned to remove the wires. Due to his continued lameness and discomfort, however, Luca was prescribed tramadol, a pain relief medication that can be given orally at home. But as it's very bitter he wouldn't take it, even when disguised in his favourite foods.

"We then tried acupuncture, starting treatment alongside a glucosamine and chondroitin sulphate supplement (Seraquin®) once a day, and pain relief in the form of buprenorphine twice a day for one week, after which it was reduced and eventually stopped. Luca was much more active at home and had stopped over-grooming his left leg. Even though he tolerated the acupuncture well initially (with food rewards), after the first 5-6 sessions (given twice weekly) he started to become grumpy when the needles were inserted, so we stopped the acupuncture treatment after three weeks. We stopped the Seraquin®, too, as Luca refused to take it.

"It is difficult to know precisely which therapy was most beneficial to Luca, but we think the combined benefits of all the treatments reduced his pain so that he became more comfortable. We tried to stop the NSAID therapy four weeks after the end of his acupuncture sessions, but Luca became lame again, indicating that he has chronic OA pain. Fortunately, the lameness resolved when NSAID therapy was reinstated. Over the last 12 months, Luca has remained comfortable on meloxicam, and his dose has been reduced gradually so that he is now on half his initial dose every day. His fur has completely grown back where he was over-grooming, and he is agile at home."

Case history: Emelia

Emelia, who is 16, is the first cat to undergo stem cell therapy, performed by Dr Jim Rummel, at Camboro Veterinary Hospital, in Pennsylvania, USA.

Emelia had severe osteoarthritis in her hips, front legs and lower back, which was restricting her formerly active life; even climbing into her litter tray was painful for her. While the anti-inflammatory medicine she was taking helped to control the pain in her joints, her owner, Fran, believed SCT might bring more relief and help Emelia get back to being a healthy, happy cat.

The treatment involved removing stem cells from the fat in Emelia's neck, activating them with laser light, and then injecting the cells back into Emelia's joints to provide pain relief and to build new, normal cartilage in her joints.

Emelia responded well and is happy at home.

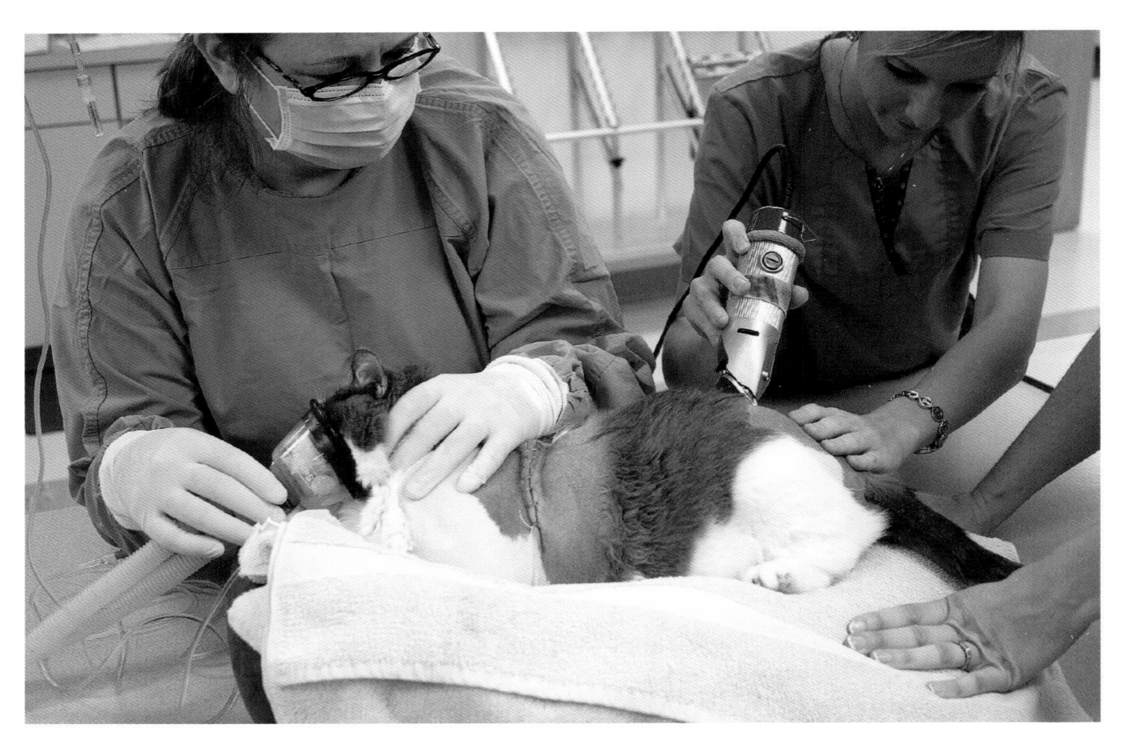

Emelia being prepared for the operation.

Chapter 5

Are certain breeds more susceptible to OA?

While any cat – purebred or moggie – has the potential to develop primary osteoarthritis (OA), the older the feline and the more weight he is carrying around, the greater the risk of this happening. The genes inherited from parents can have a bearing, too.

Size and sex considerations

Larger breeds and medium-sized ones are more susceptible to joint problems as they're generally heavier. And as boys are often fatter, tend to jump around more, and often take extra risks which could lead to an injury or some kind of trauma to a joint, they're more likely than girls to develop osteoarthritis.

The Maine Coon breed, one of the oldest natural breeds in North America, and increasingly popular in the UK, is one of the largest of the domestic cats.

Hip dysplasia (HD)

Conditions such as hip dysplasia, an abnormal development of the hip joints (see chapter 2), which can lead to osteoarthritis as a cat ages unless the damaged hip is replaced, tend to affect large-framed cats at any age.

Larger cats are more prone to OA. (Top: courtesy Teresa Labiak)

Whatever the breed, the older a cat gets, the greater the risk of primary OA.

Breeds affected include the Maine Coon, Devon Rex, Persian and Siamese.

Other developmental hip problems can affect young, heavy, male cats, particularly those neutered at an early age.

Burmese cats

Some Burmese cats can suffer from osteochondromatosis, a degenerative disease of the joint capsule, which causes them to move with their elbows held outward – a stance sometimes described as 'clockwork kitty syndrome.'

Manx cats

Manx cats carrying the genetic mutation Manx Syndrome suffer from developmental problems in their spines – similar to spina bifida in people – and can suffer arthritic changes in their tails.

Bengal cats

This increasingly popular breed has been susceptible to OA and patella luxation, a dislocation of the knee cap which

Bengals are descended from snow leopards. (Courtesy Banham Zoo)

causes secondary OA. Also affected are Abyssinians and the Devon Rex.

Barrie Alger-Street, a leading breeder of Bengals, had a cat with the wonderful

Seraphina Typha Zolustre.

name of Calusa Typhasnowstorm (Storm), who developed osteoarthritis before he died at the grand age of almost 17.

Happily, Storm's granddaughter, Seraphina, didn't show any obvious signs of painful joint disease as she aged.

Scottish Folds disease

Scottish Folds cats can suffer from osteochondrodysplasia; where the cartilage doesn't form properly, resulting in cats with forward-folded ears. It's passed on by a dominant genetic trait, which means that if one parent has genes for straight ears and the other for folded ears, the kittens will be fold. Affected cats will eventually suffer serious joint disease and debilitating pain, making them lame and reluctant to jump.

Opening up the gene pool

A pedigree cat's gene pool is often restricted, and is by no means as diverse as it needs to be to keep a purebred cat as free as possible from inherited diseases.

Some genetic experts believe the best way to keep a pedigree gene pool as varied as possible is not to breed too often from the same generation, and to wait at least five years before breeding from the next generation.

Choosing the right breeder

If you're considering giving a home to a pedigree cat, it's advisable to buy from a reputable breeder. They will be able to tell you more about your cat's background,

continued page 36

Case history: Sam

Sam has been re-homed by the RSPCA centre in Norwich, as Animal Care Manager Claire explains –

"Sam was a 12-year-old Maine Coon with diabetes, and was extremely overweight when he was signed over to us in November 2012. His owners couldn't afford, or cope with, the treatment Sam needed. We didn't notice any potential problems relating to his joints whilst he was in our cattery, as he didn't do very much, and staff assumed he was a bit of a lazy chap! However, when he went into a foster home it became apparent that it was uncomfortable for him to get around: he found climbing stairs very difficult, for example.

"On examination by a vet, his joints seemed very painful. The vet strongly suspected arthritis, but was unable to say for certain without x-rays. However, a general anaesthetic was not recommended due to the increased risk of Sam being diabetic, so we went with the vet's suspected diagnosis.

"Sam was put on meloxicam to manage his pain, and he improved quite quickly, and was able to get around much more easily. We saw a difference in his personality in that he was much more willing to walk around, jump, and climb, which proved beneficial in getting him to lose some weight! Sam was a huge character (in more ways than one!), and very popular with staff and volunteers here.

"Due to his having two medical conditions, we launched a special appeal for him online to find him a suitable home. A lady from London with past experience of diabetic cats got in touch with us, and subsequently adopted him. Sam's now doing really well. He's off his insulin and officially in diabetic remission!

"His new owner keeps in touch, and has told us that Sam's arthritis seems to play up in damp weather, but he has pain medication when he needs it, and is taking some super-duper cat glucosamine and chondroitin supplement as well, which really seems to help."

Case history: Tama

Tama, who's a three-year-old neutered Russian Blue, has had a hip replacement recently, as Dr Richard Gowan of the Cat Clinic in Melbourne, Australia explains –

"Tama lives indoors with four other cats. His owners had noticed he had become intermittently lame on his left hind limb, and was reluctant to jump. A physical examination revealed he was in remarkable health other than that he resented his left hip being extended or rotated. X-rays showed that Tama had suffered a fracture to a growth plate in the head of the femur (thigh bone) in the hip. He'd probably had this for some time.

"It's unclear what caused this fracture, but larger male cats – and especially those neutered at a very early age – have been shown to have an increased risk of growth plate fractures of the hip joint. Some kind of trauma could have played a part, too.

"As Tama is a large cat, there was a probability that the problem would occur in the other hip also. Tama's owners agreed that a joint replacement was the best course of action, as this facilitates a speedier return to full limb function, and likely a longer, pain-free joint life span, as well as a reduced likelihood of arthritis developing. A specialist veterinary surgeon, Dr Guy Yates, performed the operation at a referral hospital in Melbourne: thought to be the first of its kind in a cat in Victoria, and maybe even in Australia.

"Tama made a rapid recovery from his surgery, and started using his leg very quickly; almost using his leg normally within four weeks. His owners were thrilled. His long-term follow-up has shown Tama's hip displays normal and pain-free ranges of movement, and his activity levels are the same as they were before his injury."

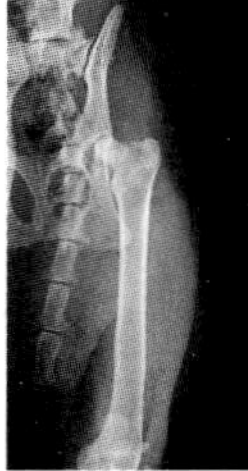

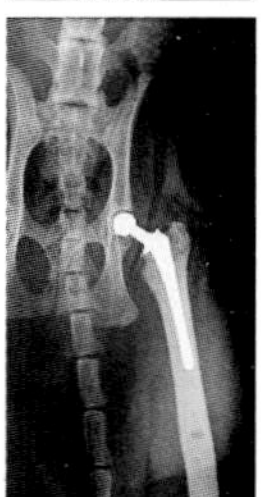

'Before' and 'after' x-rays of Tama's hip.

pedigree and history, than if you were to acquire the animal from a possibly less reliable surce.

Organisations that can help

The Governing Council of the Cat Fancy (GCCF) – the feline equivalent of the Kennel Club – is the premier registration body for breeding and showing pedigree cats in the UK, and actively promotes genetic and other relevant health tests; maintains a genetic register, and advises breeders on best breeding policy. The Council is also happy to help owners with general advice, whether their animal is a top class pedigree or a humble moggie, and can provide

Tama.

a list of accredited breeders to anyone considering getting a cat, whatever the breed.

The International Cat Association (TICA) can also provide information on breeds, and has a genetically-based registry. It holds cat shows around the world.

The University of Sydney's Faculty of Veterinary Diseases' website has further information on different breeds and inherited disorders.

Chapter 6

Diet and supplements

Cats are catching up with dogs and humans in the obesity stakes as a result of too much of the 'wrong' food, or too much food, full stop, coupled with not taking enough exercise to offset those extra calories.

Your cat needs to be lean and keen to cope with osteoarthritis, so being overweight won't help at all by putting extra pressure on already painful joints, and possibly making the disease more difficult to treat.

The importance of weight control

An estimated 25 per cent of cats in the UK are overweight, and if your cat's carrying around extra weight, it's time to take action.

As any vet will tell you, the best favour you can do your cat, particularly one with joint problems, is to diet him. Dieting can obviate the need for surgery for some cats – a good enough argument, if ever there was one, for helping your cat shed those extra pounds.

Cats are clean animals by nature, and if they can't reach certain areas of their body to groom because they're so overweight, they'll often stop cleaning themselves, which might cause them more distress than having to deal with the arthritis itself.

Temptation

It's best not to leave food lying around the house in full view of your cat, as the temptation to scoff the lot might be just too hard for her to resist.

And while cats are very good at wrapping themselves around your legs, purring and meowing for food, try not to give into the pressure with the excuse that it will make him (or, perhaps, you?) happy. But if you simply can't resist giving a treat, aim for a high protein option: pizza and cakes should definitely be off the menu for any cat.

Not too much too soon

If your cat's overweight, she'll benefit from carefully controlled weight loss, supervised by your vet, who'll probably recommend a special diet to help achieve this safely and effectively. It's not good for your cat to lose too much weight too soon: aim for around 1-2 per cent of overall body weight per week (40-80g (1.4-2.8oz) for a 4kg (8.8lb) cat). It should take around 3-6 months to lose 1kg (2.2lb), although it could be longer.

Regular grooming is an essential part of a cat's life – occasionally involving some amazing contortions!

Hendrix has to work hard for his treats!

Will certain foods help with arthritis?

Whether some foods can have a positive affect on the symptoms of arthritis isn't always clear, but certainly a healthy, balanced diet can only help your cat to deal with the effects of the disease, and maintain overall fitness.

It can be difficult to know what to give your cat, though, as what's good for people, or dogs, nutritionally-speaking, won't work for your cat, and could even cause health problems, such as kidney disease or cystitis, later on in your cat's life.

A good balance

An under-nourished cat is as unhealthy as an over-fed one, so try to make sure that your cat is getting a balance of protein (such as chicken, or fish occasionally); fat for energy; essential fatty acids to control inflammation, and vitamins and minerals, such as calcium for bone health.

The importance of staying hydrated

As cats have an inherently low thirst drive, the water they take in from food is vital. Studies have shown that when a cat eats canned food, coupled with having access to a water bowl, he'll consume double the amount of water compared to if he had a dry food diet.

Water is critical for urinary tract health, and for keeping feline joints hydrated. It's also vital for weight loss, as a cat on a water-rich diet will feel fuller, and therefore require fewer calories. It cannot be over-emphasised just how important water is for your cat's overall body health.

Wet versus dry food

Domestic cats are just like big cats in the wild when it comes to their diet, and we should always ensure that they get the right food for their species, advises US vet Dr Pierson of catinfo.org –

"In their natural setting, cats – whose unique biology makes them true (obligate) carnivores – would not consume the high level of carbohydrates (grains, potatoes, peas, etc) found in dry foods (and some canned foods) that we routinely feed them. Dry food is lacking in critical water content, and it's also typically high in carbohydrates.

"In the wild your cat would be eating a high protein, high moisture meat/organ-based diet, with a moderate level of fat, with only approximately 1-2 per cent of

A lion has no trouble climbing a tree to reach meat ...
... and a domestic cat copies his big cousin. Stretching helps their joints, too. (Courtesy Graham Law, University of Glasgow)

Diet-wise, little cats need to emulate their big brothers in order to thrive (taken at Amazona Zoo).

her diet consisting of carbohydrates. You would never see a wild cat chasing down a herd of biscuits running across the plains of Africa, or dehydrating her mouse and topping it off with corn meal soufflé."

Introducing a different diet

If you do decide to change your cat's diet from dry to canned or wet food, it's a good idea to introduce this new regime gradually. Be patient, as it might take time to convince your cat that this is a change worth making. (There are tips on how to make the switch from dry to canned food on Dr Pierson's website.) And always check with your vet before switching your cat to a different, or special, diet in case there are any medical reasons not to. Most veterinary practices have a nurse specialising in nutrition if you need further advice.

A homemade diet

If you're happy to prepare food for your cat at home – and have the time to do so – this could be the way to go, as long as the correct balance of protein, fats, vitamins and minerals is achieved.

Says Dr Pierson: "One common mistake people make when feeding a home-prepared diet is the belief that a cat can live on meat alone, without bones as a source of calcium. While meat must be the primary component of a feline diet, there is not enough calcium in meat (without the bones) to provide a proper calcium-to-phosphorus ratio. Always remember that calcium is not an optional 'supplement' but a very critical component of the diet."

Dietary supplements

Owners like adding supplements to their cat's diet to help with arthritis, in preference to giving them strong drugs. Yet there's no scientific evidence that supplements work. As arthritis symptoms fluctuate, it's hard to tell if the supplements are really helping, or your cat's just going through a good phase.

Always check with your vet before giving your cat supplements, as some might react with prescribed drugs, and can even be dangerous if taken in high doses. And ask your vet for advice on which they would recommend trying.

You might feel tempted to give your cat the same supplements that you take, but this isn't advisable, even if it seems the cheaper option. There are a number of supplements on the market, and sometimes it's hard to know which to go for. Supplements containing glucosamine and chondroitin, such as Seraquin®, are among the most popular ones for cats.

As supplements can be expensive to purchase in the conventional way, buying them online should help keep costs down.

Case history: Cookie

Cookie, 16, from Middlesborough, took part in a national diet and fitness programme run by the PDSA, the UK's leading veterinary charity. At the start of the six month programme, she weighed in at a staggering 8.4kg (19lb), making her around 90 per cent overweight. Cookie's now a trimmer 7.3kg (16lb) after dropping 1.1kg (2.4lb) – equivalent to 13 per cent of her total body weight.

Her body shape has changed, too, and she's lost 17cm (7 inches) from her waist and the same from her chest. While there's still some way to go, Cookie's diet, under guidance from the PDSA, is clearly making a difference.

Cookie was rescued as a kitten by her owner, Laura. Says Laura –

"When the vet told us how overweight Cookie was we were determined to do something about it, as she's an old lady now and we want the best for her. She never begged from the table but we did give her titbits and top up her biscuits when the bowl was empty. She was also a food thief – we once caught her tucking into roast chicken that had been left to cool in a neighbour's house!

"Cookie took to her new diet very well, although there were a few hiccups along the way: mainly my dad not being able to resist Cookie's pleading miaows and giving her a sneaky piece of ham or chicken.

"Cookie is much healthier and happier now – and a lot more vocal than she used to be! Previously, she would sleep most of the day, but since starting Pet Fit Club, she's definitely more active."

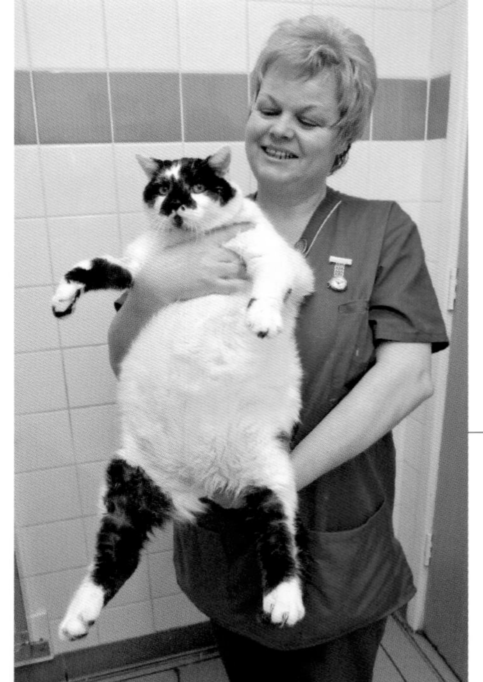

Cookie before her diet, with PDSA head nurse Steph ...

... and after. Cookie lost the equivalent of a bag of sugar.

Case history: Willow

Says Laura, Willow's owner –

"Willow was diagnosed with arthritis when he was eight-and-a-half. Shortly after we found out he had chronic renal failure (CRF), which meant that, unfortunately, he wasn't able to have meloxicam for the arthritis.

"I noticed he was weaker on his back legs, and was hardly able to move sometimes. I really did feel for him, and knew I had to do something for him. His vet suggested we try glucosamine. I thought it couldn't do any harm, and it was safe for him to use with his current treatment for CRF.

"I didn't see any benefit for about six weeks, and was growing impatient that it wasn't working. Willow was happy, though, and his CRF wasn't bothering him too much. Then, one day, I was sitting in the garden and he leapt up the six foot high fence in our garden – something he had not been able to do for a while. I had a huge smile on my face, and can't begin to tell you the elation I felt for him!

"I had my Willow back and, sure enough, he was jumping and leaping about all over the place; even playing fetch and kill with his favourite Kong® toy, bunny-kicking it. On his next appointment at the vet I couldn't wait to tell them the news, and they were just as happy as I was. We continued the glucosamine treatment with one tablet every day.

"During the winter months Willow did slow down a bit, but he was a fighter and never ever showed any pain or discomfort. I used to wrap a hot water bottle in a fleece for him, and he would curl up and fall fast asleep on it. I also massaged his legs to ease any pain or discomfort he may be experiencing.

"During his last few weeks with us, I knew he was growing frailer, and his CRF was taking a toll on him, but he was still able to walk and jump, right up to his last morning when his body gave up. That extra year-and-a-half of taking the glucosamine really did help him, and his quality of life improved. To see the difference in him was extraordinary, and all without pain relief. I would recommend glucosamine to any owner with a cat suffering from arthritis, and especially a CRF cat with arthritis."

Willow benefited from glucosamine.

Chapter 7

Complementary therapies

Complementary therapies can offer relief from the symptoms of osteoarthritis, or some of the side effects of conventional drug treatments. And a number of owners believe their cat's life can be extended by a few years with the help of these therapies.

Generally speaking, complementary therapies are safe if practised by a qualified therapist from a recognised organisation: a referral from your vet is usually needed.

Holistic approach

While conventional medicines focus on treating the pain and inflammation that goes hand-in-hand with arthritis, complementary therapies take into account how the whole body is functioning (hence holistic), including lifestyle, personality, and diet. Complementary therapies are designed to be used alongside conventional medicines and treatments, and not to replace them, although some owners find their cats live happily on these therapies alone.

Are they effective?

While scientific research is still relatively recent, and currently small-scale, early results indicate that some therapies may indeed help ease the symptoms of arthritis, although they can't alter the course of the condition or provide a cure, sadly. Sometimes it's the case that two different therapies can work well together.

Acupuncture

Based on ancient Chinese theories about how energy moves around the body, this therapy uses fine needles which are inserted through the skin at certain points of the body, and left in for around 30 minutes usually, to help relieve pain and promote general wellbeing. Acupuncture can be slightly uncomfortable for your cat when the needles are inserted, but shouldn't hurt her; some even doze off during a session. Acupuncture can only be performed by a vet, although acupressure (using fingertips instead of needles) can be practised by anyone qualified to do so.

Homeopathy

Homeopathy is based on the principle of 'like cures like.' Where conventional medicine aims to suppress symptoms – for example, by using anti-inflammatories

– homeopathy provokes the body into healing itself. Homeopathic remedies like Rhus Tox can help cats with problems such as stiffness that eases on movement.

Massage

This therapy treats the whole of a cat's body to reduce soreness, tension, over-compensation, and old muscular injuries. By manipulating muscles and restoring their function, massage is a natural way of managing your cat's health, prolonging quality of life, and promoting wellbeing. It is of particular help to cats with hip dysplasia.

Osteopathy

The principle of osteopathy is to restore normal function throughout the body, and reduce tension in areas where muscles are too tight, whilst ensuring that other regions are 'woken up' to spread the load across the body. Osteopathy uses gentle manual techniques and soft tissue massage to improve flexibility and range of movement. Osteopaths tailor their treatment to best suit each patient. The treatment will often result in a relaxed cat, who is brighter and more willing and able to exercise.

Gentle stroking is also relaxing for a cat, and can help reduce any tension around the joints.

Case history: Boris

A 13-year-old Russian Blue owned by Carole –

"Around two years ago Boris was diagnosed by his vet – and confirmed by x-ray – as having osteoarthritis in his hips and elbows. Looking back, he had probably had it for much longer than that as he has changed from being a formidable hunter, often bringing 'presents' of mice and birds, including large magpies, through the cat flap, to a cat who rarely goes outside. Like most tom cats he also used to like to roam, often far away, but gradually became less mobile.

"This change happened over a number of years, and was only really obvious when his mobility became slightly laboured.

"Boris' vet prescribed meloxicam to reduce the pain and improve his mobility, and he still takes this daily with his food. However, Boris was still in considerable discomfort despite the medication, so his vet suggested trying acupuncture, and Boris was referred to another vet who specialises in pain management.

"Boris initially had weekly treatments, followed by monthly visits from then on. He tolerated the needles very well, and barely flinched throughout the treatment: in fact, he seemed to find it strangely relaxing, and would pummel his blanket throughout the treatment. On the way home from each of his treatments he would sleep with his paws stretched out like superman.

"Following a few sessions there was a noticeable improvement in his mobility. He began sleeping on my son's cabin bed again, which he'd been unable to get on to as it's high up. He also began using his cat flap again."

Boris received treatment from Samantha Lindley, Chronic Pain Clinic, Glasgow Vet School.

Boris at home.

Case history: Chloe

13-years-old and owned by Pat –

"Chloe hadn't been eating well on occasion, and had lost some weight. We hoped that by having her teeth cleaned, the problem would be resolved, but, unfortunately, it wasn't.

"It was then decided to have some blood tests carried out which showed that Chloe had arthritis, and the pain of it (which was not obvious to us) was putting her off her food.

"I was offered meloxicam, but I was not happy about giving this drug to Chloe. As I prefer to treat myself in a natural way whenever possible, I thought, why should I not take the same view with our cat? After looking into homeopathy (my favourite therapy) Nick Thompson, the Holistic Vet, came highly recommended. After the consultation, we were able to start Chloe on a remedy. Initially, it was trial and error with the best way to give the drops to her, and the most successful method was to put a drop of the liquid on my clean finger, insert the finger into the side of her mouth and rub it along her gum.

"I then discovered that one of the vets at our practice did acupuncture. Chloe began a course and, amazingly, would sit on the vet's examination table, where she could also look out of the window, while the vet, speaking gently to her, put several needles along her back to the start of her tail. Sometimes, she would lie down on the table, completely happy and relaxed – and would even purr when I spoke to her!

"Chloe continued to have both treatments alongside each other for a few weeks, and for a while she was symptom-free: eating well, with good movement. Then I noticed her right back leg was not straightening as much as her left back leg – particularly when she was being stroked – and so I resumed giving one drop of the remedy at bedtime and in the morning her appetite was returning to normal, and when I opened the back door, she ran out.

"At a recent check-up our vet found that the muscles at her rear end were very tight. Consequently, he carried out an acupuncture treatment and her muscles softened, and she is, once again, a happy, pain-free girl with a good appetite."

Chloe's found her feet again.

Case history: Hamish

Hamish is being treated by leading osteopath Tony Nevin of Zoo Ost Ltd, which is based in Gloucestershire –

"When he was nine years old, Hamish was injured in a road accident, and had surgery to pin his left hind leg. Although the bony tissue had healed nicely, his owner, Meg, noticed that, consequently, he walked with a limp.

"When I met Hamish, some five years after the accident, he had a definite list to the affected side (left), and some muscle wastage from behind his rib cage, along the spine and over his rump, which suggested trauma had also affected his spine.

"Although his injuries dated back over a third of his life, Hamish was less than happy for anyone to feel around his back end and along the base of his spine: as he loved to be made a fuss of around his head and ears, this is where I started the first session. Gently altering the position of his head in relation to his neck and shoulders had the desired effect of altering tension through his back end.

"I was then able to move to his neck and throat, all of the time going at a pace that he liked: this was easy to monitor as Hamish is very vocal and only too willing to let me know if he didn't like anything. I worked slowly on various parts of his body until both he and I felt enough had been achieved in the first session. Meg's homework was to try and ensure that Hamish avoided obstacles too challenging for him before his next visit – including trying to get over the garden fence to be let in, if Meg hadn't heard him.

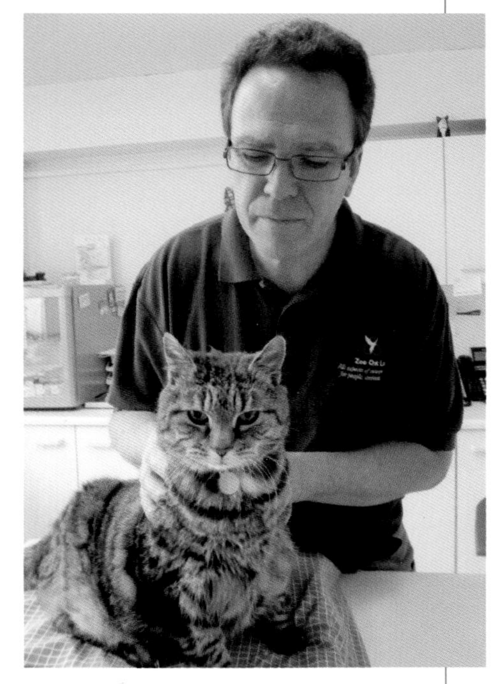

Hamish receiving treatment from Tony.

"Cats are somewhat harder to treat than dogs as they are rather strong-willed and mostly more independent. Where you can win them over is in making sure they have lots of pampering, so Meg and I discussed a sensible plan for Hamish, and he was then booked in to see me a week later. I knew he'd need a few treatments close together to get him to age more evenly throughout his body, and to be as comfortable, and as symptom-free as possible. Over the next few weeks I saw him weekly, then fortnightly, and then monthly. I now see him roughly every two to three months, depending on what he's been up to.

"On the whole he looks much better, and is even willing to let me touch any part of his body without showing any signs of discomfort. Meg says he's a much happier member of the family, and that – coupled with his acceptance at being gently stroked and felt with a flat hand – are sure signs that Hamish is improving."

Chapter 8

Exercise and physiotherapy

Regular exercise is good for your cat as it nourishes joint cartilage, and will help to keep her arthritic joints from stiffening and becoming more painful. Ultimately, exercise goes a long way toward keeping your cat on the move.

Playtime

Whilst cats aren't the most dextrous of animals, they do love to play, and gentle games are a good way to keep your cat active in mind and body; hopefully without injury. If you have another cat at home, the two will probably play together happily, inside the house or out in the garden: single cats are likely to look to you for entertainment. If your cat usually ventures outside but seems reluctant to now, you could try putting his food outdoors: the smell may encourage him into the garden to investigate.

Interactive games

Try to think up games you can play together. Simple activities like dragging a furry toy across the floor on the end of a piece of string will encourage your cat's natural predatory behaviour. And rolling a ball for your cat to chase after will get her moving, and might keep her amused for a while – until she decides to call time, of course. My cat never tired of playing 'jump on the slithering snake,' in the shape of one of my belts that I swished enticingly across the floor.

Stretching

A tall scratching post will encourage your cat to stretch, which will improve flexibility and helps her joints move through their full range of motion.

The importance of exercise

It cannot be overstated just how important exercise is in reducing the secondary effects – such as stiffness and weakness – of the pain and inflammation associated with arthritis that your cat will be experiencing. Exercise will also help to keep her weight under control to avoid further pressure on her sore joints. If you're worried that your cat isn't sufficiently mobile, you could ask your vet, or physiotherapist, to suggest ways to help your cat to be more active.

Veterinary physiotherapy

Whilst it's widely known that arthritic

Cats naturally love to play.

or injured dogs can benefit from physiotherapy, a view seems to prevail that cats won't tolerate the handling needed to benefit from this treatment. Yet there's huge potential in physio for cats. And with modifications and adjustments – and the use of the correct techniques – physio can significantly help a cat who's suffering from arthritis following surgery, perhaps for a hip replacement, after an injury, or because of normal wear and tear. Cats with inflammatory arthritis can be helped, too, but with more targeted, delicate treatment.

Lilly-Pie, the Maine Coon, enjoys stretching. (Courtesy Josie Hughes, breeder)

Sadly, although physiotherapy can't reverse your cat's arthritis, it can relieve some of the symptoms, and, with careful management, slow disease progression. It can also work well alongside hydrotherapy (see chapter 9).

The benefits

Physiotherapy is concerned with movement and physical function, and the aim of treatment is to return an animal to as near full function as possible. This can be achieved through strengthening muscle condition; improving flexibility; restoring balance, and controlling pain and inflammation.

Physio can also help in the management of chronic pain associated with joint problems, and in restoring mobility.

Referral process

In most cases, a referral by a vet is needed before your cat can begin a course of physio. Your vet will have a clear picture of your cat's arthritis with which to brief the therapist, and will probably know the best veterinary physiotherapist in your area for your cat's particular needs.

An initial assessment

The kind of arthritis your cat's suffering from, the condition of her muscle mass, and her general fitness will be assessed at the first session.

The physiotherapist will consider your cat's body as a whole: how it functions, and its mobility. How your cat stands, walks and sits will be assessed, too, and whether (and where) there's stiffness or lameness. If your cat's trying to compensate by shifting weight from one leg to another or leaning to one side, this could indicate problems.

And you're likely to be asked, among other things, where your cat sleeps – on the sofa? On your bed? Upstairs or downstairs? – and how active your cat is.

The physio will also ask about what medication your cat is taking; if his behaviour has changed recently, and what you, as his carer, expects from the treatment.

The approach

As cats might not tolerate a lot of handling, treatment sessions will probably be short, and will utilise your cat's natural love of play – ideally, with familiar toys from home.

Techniques that can help

Physiotherapists have a range of techniques to call on to reduce pain and inflammation, and get your cat back on her feet, and these usually include massage to improve circulation and muscle tone, reduce swelling, and break down adhesions (scar-like tissue) following injury. Also among the options are passive stretching techniques to encourage your cat to flex her joints and restore movement as soon as possible. These exercises can be demonstrated to you to continue doing at home; ideally, twice a day.

Other treatments

Sometimes – and particularly after surgery – your physio might apply a cold wrap around the affected limb (cryotherapy) to control pain, inflammation and swelling. Heat therapy (thermotherapy) might then be used some days later to aid healing, and improve the effectiveness of stretching exercises. A therapeutic laser, or other form of electrotherapy, might also be used for inflammatory and soft tissue injuries.

Timescale

It can take from three to six months to reduce your cat's joint discomfort, and build muscle mass and general fitness.

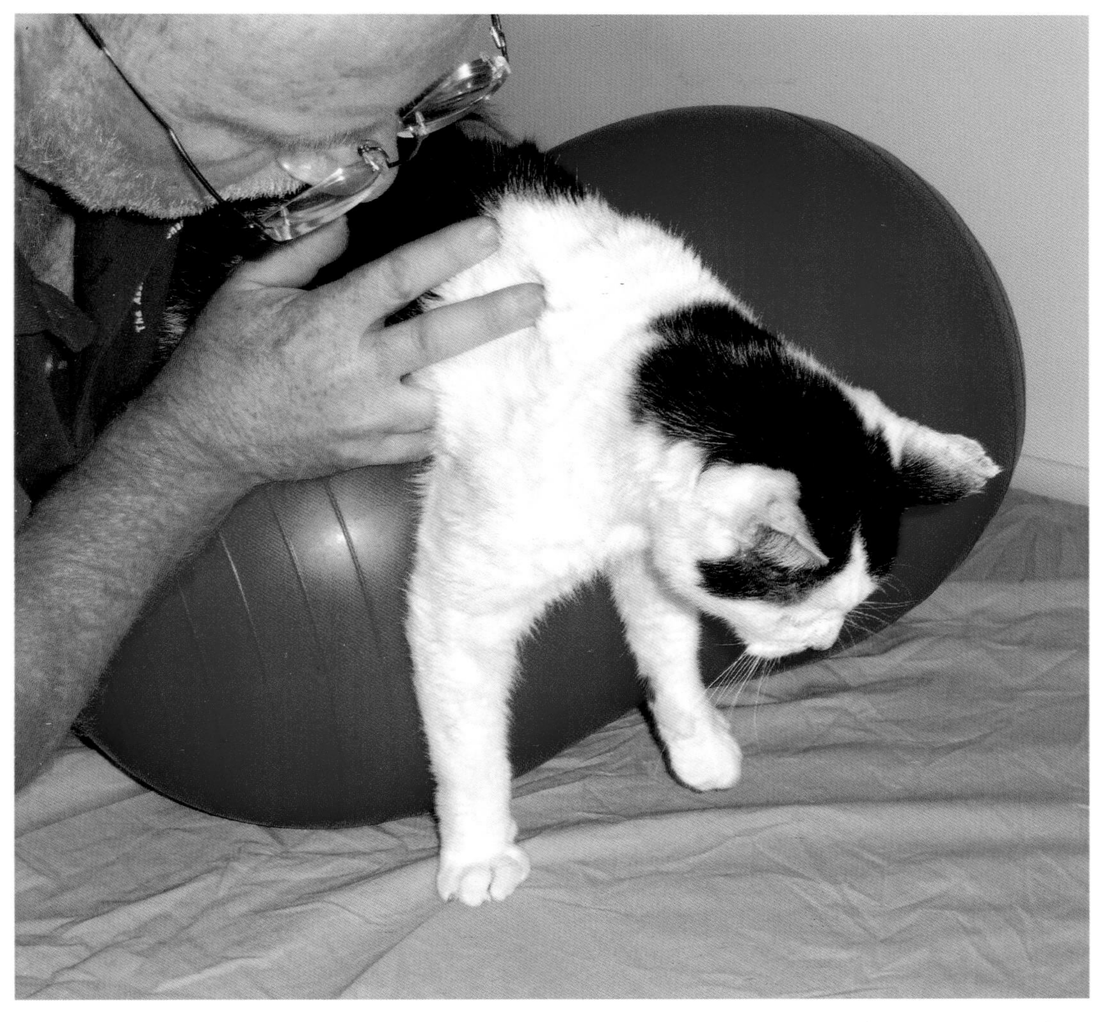

The use of a physio-roll provides sufficient support to allow a cat to perform exercises that might not otherwise be possible. (Courtesy Brian Sharp of Canine Physio)

For the first two to three months you might see the physiotherapist weekly, and then the visits will probably gradually reduce to once every three to four weeks on average.

Working together

Your therapist should explain fully why your cat has become arthritic, and why each treatment is important at each stage. Feedback from you – is your cat better or whether there's been a flare-up of symptoms, for example – will help make the rehabilitation successful.

Case history: Bert

A British Blue owned by Kate –

"We were always aware that Bert had a slightly unusual way of standing, but when he was approaching two years old, he began to experience problems with jumping up on things, and seemed to stop grooming himself around his rear end. We discussed this with the vet, who suspected hip dysplasia, which x-rays confirmed. Bert had surgery on his left hip just before his second birthday in 2010, and then further surgery to remove a bone spike.

"A few months on and Bert had improved physically, although the painkillers weren't helping, and he seemed so miserable. This is when we decided that we needed something more or the only alternative would be to have Bert put to sleep to end his suffering. We began looking at the alternatives, and physiotherapy was suggested, as it had proven very successful in helping dogs post-operatively.

"We were uncertain how Bert would react to going somewhere for physio, as he was now very fearful of the vet. But as a last option we were willing to give it a try. Happily, Bert settled in to the environment really well, and seemed to enjoy the physio sessions. He was given laser therapy and exercises at his sessions, and we continued with his exercises at home: my husband made smaller versions of the equipment so we could keep up the work.

"We religiously did his exercises three times a day, and began to see an improvement. The day Bert was signed off from physio we were very happy: physio had essentially turned his life around and made him pain-free. To this day he lives a happy life climbing and jumping, and living life as a normal cat."

Bert was treated by veterinary physiotherapist Sarah Clemson at Meadow Farm Rehabilitation & Hydrotherapy Centre in Norfolk. Sarah says –

"Usually, cats recover well following this type of surgery, but the pain caused by the bone spike had delayed Bert's return to function. To further complicate his situation we suspected that, after the second operation, his sciatic nerve had become irritated or impinged against the bone in his pelvis, inhibiting him from using his leg normally.

"A tailor-made exercise programme was created for Bert, with the goals of reducing his pain; increasing weight-bearing ability; building hind limb muscle bulk, and optimising overall function and wellbeing. His exercises included balance work on the wobble cushion, and pole walking to increase his strength and range of motion in his hip.

"Bert responded well to therapy, and as he began to feel more comfortable he started to put more weight through the leg that was operated on, gradually building muscle bulk and being able to walk further without becoming lame.

"Over an eight-week period of treatment and 'functional' graduated exercise, Bert was able to begin resuming normal activities. He practised climbing the stairs again, and began going outside for short periods. We knew he had residual physical restrictions, but cats are incredibly resilient, and we hoped that, once he got back to doing his usual activities, he would continue to improve and learn to function well within his limitations."

Bert doing his proprioceptive exercises on a wobble cushion.

Bert pole exercising.

Chapter Nine

Hydrotherapy

Cats and water are not supposed to mix – that's the widely-held view, at least.

Certainly some cats would do anything to keep their paws dry, yet certain breeds – such as the Turkish Van (nicknamed the swimming cat) – and Bengals, Burmese, Siamese, Persians and Norwegian Forest Cats often take to the water ... well, rather like a duck would.

Naturally curious cats will probably respond well to hydrotherapy, particularly if there's a treat waiting at the end, but the therapy might not be suitable for cats who have a tendency to bite or scratch when stressed or frightened.

Your cat would have to co-operate, and the stress of being in the water should be weighed against the benefits of having the treatment.

Whether your cat loves water or has a take-it-or-leave-it approach, hydrotherapy has definite benefits, and is certainly worth a try.

It's a safe, all-round exercise option that will help to manage your cat's arthritis without putting his poor old joints under further strain, and improve his mobility at the same time.

Land versus water

Each time your cat moves on land a shock wave travels up her limbs and is absorbed by bones, tendons and joints. While weight-bearing exercise helps to maintain healthy, strong bone, if exercise is severe, repeated too often, or is high impact, it could damage or weaken an arthritic joint, or one recovering from injury or surgery.

With hydrotherapy, the buoyancy of the water bears the load, although the muscles still have to work hard; harder, even, than they do on land because of the resistance of the water.

Water-based exercise uses around 30 per cent more oxygen than similar land-based activity, and cats often go faster in the pool or on the treadmill than they would do on land.

The benefits

The advantages of a few hydrotherapy sessions include: strengthening and toning most muscles; increasing muscle bulk, and relieving joint pain, swelling and stiffness. It's a good, all-round cardiovascular experience for your cat, which can only help her cope better with arthritis, and

Zorro. (Courtesy Cleasanta Norwegian Forest Cats)

increase joint flexibility. And chances are your cat will feel more confident back on land, too, which is a bonus.

Reversing muscle wastage

As feline muscles can begin wasting within just a few days of becoming immobile, it's important to prevent any further weakness – and possible injury – by re-building those deteriorating muscles as soon as possible through safe exercise such as hydrotherapy.

Treadmill

Most cats receiving hydro will wear a life jacket or harness, and exercise on a treadmill or water walker, although some are happy to swim in a pool. The treadmill has glass sides which allow the therapist to monitor a cat's gait, range of movement, and length of stride. (If in the pool, the therapist will usually be in there, too.)

As the buoyancy of the water takes the load, reducing weight on a painful joint, or one that's healing, your cat is more likely to stretch his legs further in the water – and might then be happier doing this when out of the water, too.

Improving circulation

Warm water increases blood circulation to the muscles, helping them to relax, and boosts the supply of oxygen and nutrients

Kovu having a session at Hawksmoor Hydrotherapy.

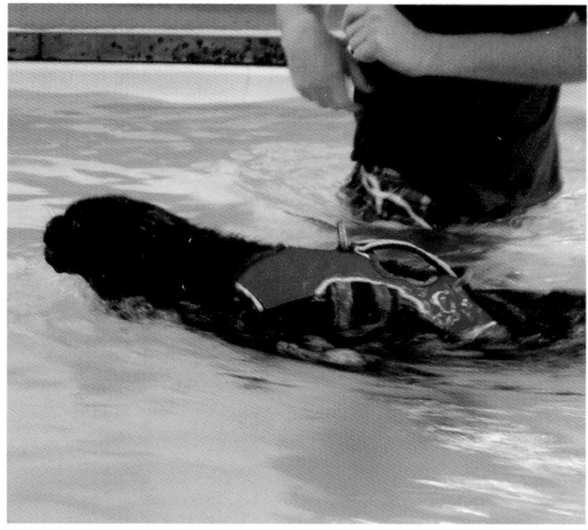

Joe in his swimming gear, and in the pool. (Courtesy Blossoms Hydrotherapy)

to the tissues. It also helps to flush away waste products.

Mood enhancer

Apart from the physical benefits, hydro can have a positive effect on your cat's mood, and also provide mental stimulation. Cats can be quite perky when they leave the pool, and there's often an overall improvement in their mental state within a couple of weeks.

Other advantages

Hydro can help with hip and elbow dysplasia, and recovery from injury such as cranial cruciate ligament damage. In addition, alongside weight loss and a healthy diet, it is beneficial in tackling obesity.

First steps

Hydrotherapy treatment usually requires a veterinary referral to ensure there's no reason why it might not be appropriate – should your cat have an underlying condition such as diabetes, perhaps.

At the initial consultation the therapist will give your cat a head-to-tail health check: assessing his circulation; looking for any lumps or bumps, and checking the condition of his ears. Your cat's muscle mass measurements will be taken, and his gait evaluated. He'll be showered before he goes on the treadmill, and again after the session, then dried by hand, or blow-dried with a dryer.

Whilst hydrotherapists can't diagnose problems, they'll refer your cat to the vet if they suspect something's amiss.

Time in the water

How long your cat will stay on the treadmill or in the pool rather depends on why she's having hydro in the first place. Cats can go swimming up to twice a week, and time on the treadmill can be built gradually according to her ability and condition. As a general rule, cats quite like their first go on the treadmill as they have no idea what to expect; the second time might be okay, given the cat knows what she's in for. And by the third time, she'll probably have settled down, and even quite enjoy the experience. But, of course, every cat is different, so it might be a case of trial and error in the beginning.

For some cats the time on the treadmill or in the pool builds quickly: overweight felines, or those with heart murmurs, usually take it more slowly.

Ideally, your cat should be assessed before and after each swim, including checks of her temperature. (Cats suffer heat exhaustion more quickly than humans, so pool temperature is lower than that which a person could usually tolerate.)

Feedback

After the first session your therapist will need feedback from you about how your cat has responded to the treatment. Was she stiff or tired afterwards, for example, which might indicate she's overdone her first session. As owner you should go by your instinct if you feel something's wrong, do tell the therapist and they'll make adjustment.

After the tenth session your vet will be given a full report on improvements to your cat, including flexion and extension.

How many sessions?

It's important, of course, not to ask too much of your cat, but it's advisable to have twice-weekly hydro sessions for the first two weeks, as just once a week would mean that benefits would take longer to become evident. Long-term, once a week should be all your cat needs. Some pet insurance firms will pay for only ten sessions, so check your policy first, or be prepared to pay for more sessions yourself.

Trained therapists

Hydrotherapy is perhaps not as well regulated as it should be in the UK, so it's advisable to use a centre registered with the Canine Hydrotherapy Association (CHA) or NARCH, the National Association of Registered Canine Hydrotherapists. Most pet insurance companies will only pay for hydro practised by a therapist registered with one of these organisations.

Coco on the treadmill.

Case history: Coco

Coco, 16, has been enjoying her sessions on the treadmill at the Avonvale Vet Centre's hydrotherapy facility in Wellsbourne, Warwickshire, as therapist Helen Reeve explains –

"Coco's osteoarthritis was such that she was weight-bearing more on her left side to compensate for the discomfort and restriction on the right, and was limping quite badly, so her vet referred her to us for some gentle hydro. It was decided that once, or twice, weekly sessions would be of most benefit to Coco. Amazingly, Coco's introduction to the treadmill and water didn't worry or faze her at all: she got the idea immediately, and has been so good she puts many dogs to shame!

"For the first three sessions she was fitted with a life jacket, with either myself or my colleague, Emily, in the tank with her, as well as an extra body to help work the treadmill controls from the ground. After that, because of how confident she was, it was safe enough for Coco to go in on her own with just a simple harness to secure her.

"The first few sessions revealed some restriction through the right foreleg, which caused her to head-bob as the foot was placed, but that was about it. The non-weight-bearing and pain-relieving properties of the water allowed Coco to feel comfortable enough to be able to use her right leg as well as the left, as midway through the course she was looking much more even-paced, and less restricted in the movement of both forelimbs. She really gets the chance to stretch, extend, and flex all her limbs when using the treadmill without the discomfort she would have endured prior to treatment here.

"After her initial course of ten sessions, Coco's muscle mass measurements in the forelimbs were 13cm – she'd gained 2cm after starting hydro. She now has a monthly maintenance session."

Coco's owner Colette is delighted with how Coco has progressed, and has noticed a marked improvement in her activity at home since starting hydro –

"I've had her since she was a kitten, and she's had arthritis for years – it probably started after she was hit by a car. Coco has been on meloxicam, but since the hydro sessions her limp's gone and she's a lot less stiff. She runs around the garden now, chasing butterflies, and jumps up on window sills: she seems so much happier. I used to think that water and cats don't mix, so I'm glad I've been proved wrong for Coco's sake."

Chapter 10

When euthanasia is the only option

There comes a point in every cat's life when the most loving thing you can do for her is to give her a peaceful and pain-free death. Cats with severe arthritis, whose quality of life is so compromised that they can't perform basic things without pain, really need to be set free from their suffering. Deciding to end your cat's life is likely to be one of the hardest and most painful decisions you'll ever have to make.

When is the right time?

There are usually tell-tale signs that your cat is finding the pain of arthritis too much to bear. He might be eating less; appearing quieter; spending time by the radiator, or not wanting to be stroked: in essence, withdrawing. Your vet will be able to help you decide if now's the time to say goodbye.

Letting go

Most vets agree that cats, like dogs, usually decide when they want to go, but sometimes hold on because they worry their owner might not be able to cope after they've gone. If you think this might be the case with your cat, you could say something to her along the lines of: "I can see you're suffering, and I think you'd like to be set free. I'll miss you so much but I can't bear to watch you in so much pain." Your cat might then feel that with this 'permission' from those she loves the most, she can, at last, let go.

The euthanasia procedure

Euthanasia is considered a quick and peaceful death (in Greek the word means 'good death'), which usually involves an intravenous injection – an overdose of barbiturate – that typically induces death within 30 seconds. Previous to this, your vet might give your cat an injection to make her very relaxed and sleepy – almost unconscious – so that you can say a final goodbye, before the second injection that stops her heart.

A final resting place

Try to give some thought beforehand to what you would like to happen to your cat's body. This is a personal decision, and really only yours to make, so don't worry about what others might say or think.

You can bury your cat in your own garden (check with your local authority that it has no objections) although don't wrap

Norfolk Pet Crematorium's Garden of Remembrance.

his body in plastic of any kind as this will interfere with the natural decaying process. Instead, you could wrap him in a towel or his favourite blanket. Bear in mind, though, that if you move house, you would be leaving your cat behind in her final resting place.

Pet cemetery

The alternative to a home burial is in a pet cemetery, which is less personal but has the benefit of allowing you to visit the grave, even if you move house, and all arrangements are taken care of.

If you've had her cremated, you could scatter her ashes in the garden; perhaps where she liked to lie in the sun, or under the shade of a favourite tree. Or you could keep the ashes in a special container in the house, or bury this in a special place in the garden.

Coping with your loss

It's not easy to cope with the loss of your feline friend, and there's no right or wrong way to grieve. Give yourself time to mourn, and don't rebuke yourself for being upset about your cat's death. Only you know what she meant to you and for how long you'll need to grieve: there's really no time limit. Cry if you want to – for as long as you want to.

It'll help to remember the good times you had with her, and to remind yourself that she's no longer suffering.

There are various ways you can remember your cat, and thank her for all the pleasure she brought to your life. Most veterinary practices, local animal rescue centres and pet forums have an 'In Memory of' page on their website where you can place a photo and a special message for your cat – and perhaps light a virtual candle for her. And if you're finding it particularly difficult to cope, you could ask your vet to recommend a bereavement counsellor, or use one of the pet bereavement helplines.

A wooden sleeping cat, which can hold the ashes of your companion.

Case history: Tuppence, who belonged to Pam

"Tuppence came into my life one Guy Fawkes Day. He'd been running up and down the road; one of my neighbours brought him to my door, and he never left. I checked with the RSPCA to see if it could trace his owner, but he wasn't microchipped and didn't have a collar or name tag.

"One vet said he was around four or five years of age, whilst another claimed he was seven or eight. I worked on the basis he was probably nearer eight, and I had him for nearly six years, so he lived to quite a good age. He had some arthritis in his little back legs, which the vet thought could have been caused by an accident, and would probably lead to problems when he got older.

"As Tuppence aged he certainly found it difficult to go up the steep stairs in my house, although he was okay walking around. I thought maybe I could train him to ride upstairs on my mother's old stair lift but I was worried the noise might frighten him. He was on meloxicam to relieve the inflammation, and glucosamine which I put in his food.

"He was active until the end. His favourite game was playing with a comb which I'd put on the bathroom stool; he'd knock it off several times, then pick it up, walk through to the bedroom and drop it on the carpet. I'd say 'What a clever boy!' He'd often stand at the top of the stairs and miaow loudly until we played with a ping-pong ball. It didn't matter what I was doing, he always wanted to play.

"I was so concerned about his arthritis I never thought he'd have problems with his heart. He had a dry cough, so I took him to the vet, where he was given steroids and antibiotics, which helped for a short while, but I was still concerned and returned to the vet. Heart failure was diagnosed, as a result of which his breathing became difficult. It was during this visit that, sadly, Tuppence's health deteriorated very quickly, and he had to be put to sleep. I brought him home and was going to bury him in the garden, but decided instead to have him cremated.

(and left) Tuppence.

"The crematorium gave me a wooden sleeping cat to put the ashes in instead of a casket, and it's now on my dressing table. I've been very depressed since I lost him – he was such a character and I loved him dearly. I rang the Blue Cross Pet Bereavement helpline several times and talked to the counsellors, and that's helped a lot.

"A strange thing happened the night Tuppence was put to sleep. A little black and white cat came into the house, went upstairs, came back down again and jumped on my lap. She became a regular visitor, so I found out where she lived and now have adopted her. Twinkle has been a great comfort to me since I lost Tuppence."

Case history: Skye, who was owned by Alison

"Skye was three in 1997 when she was brought into Wood Green Animal Shelter, where I work. According to her owner, she was too shy, and the family wanted a more boisterous cat who liked to play with children. I didn't mind whether or not she was shy – I fell in love with her sweet nature. She was gentle and lovely and I felt sorry for her. We'd recently lost a female cat (I mainly had males), and I wanted a girl.

"Skye's shyness meant it took her quite a long time to settle and feel comfortable but we accepted her on her own terms, and she was the most ladylike cat we've owned. She tolerated my partner, and was very much a 'mummy's girl,' and only I could pick her up, or brush her.

"I've had lots of cats, and they've all been very strong: mini hooligans, really. Yet Skye (with an 'e' as I like Scotland) used to sit with her legs crossed and ate very nicely. She was such a genteel little old lady.

"She began showing signs of arthritis when she was eleven or twelve. We'd never had problems with her before: she was very healthy, although a little reluctant to exercise. I didn't want to put her on meloxicam straight away because of the damage it could do to her liver, so I had a look on the internet and bought a dietary supplement – seraquin – which I gave in her food: one tablet every other day, then one a day. It seemed to make a real difference straight away, and she walked more easily.

"After a couple of years like this I could see that Skye was getting worse. She was reluctant to jump on and off the bed – it was a lot harder for her. And after she'd been sleeping on one of her squidgy cat beds, she had to spend some time loosening up before rising. She didn't like exercise or going outside, and became a little overweight as a result – which didn't help the arthritis. The vet asked me what I would like to do and we discussed putting her on meloxicam which I had reservations about, but would, I knew, ultimately help her cope with the pain, and improve her quality of life. So she went on the medication for a year.

"Skye was still obviously in pain, however, and didn't want to get up. We had to take food to her, and she wasn't grooming herself either, though did drag herself to the litter tray. And she tried to lift her head and shoulders to greet us, but her back legs just couldn't support her. She became more and more quiet. Even so, she was still loving, affectionate and calm, but we knew the time had come. When we finally made the difficult decision to let her go, we knew she would leave a big gap in our home and in our lives. She was put to sleep at the beginning of 2012 when she was 16.

"It was a sad decision to make but we knew it was the right one. The really difficult bit was after we'd rung the vet to advise of our decision: we had to wait what seemed like ages for an appointment. I remember thinking when I was waiting: here I am looking at you, Skye, for the last time before we have to say goodbye.

"But we have lovely memories of Skye, our genteel old lady to the end. She was cremated and I have her ashes in a casket at home, so they (and the ashes of all my other cats) can be scattered with mine when I go!"

Skye. (Painting by Natalie Knowles)

Chapter 11

Conclusion

Being told your cat has arthritis is probably not something you ever want to hear. It's generally a debilitating disease which can take some adjusting to – for you and your cat.

But on a positive note, there's a lot that can be done to relieve your cat's pain and stiffness, including the right prescription drugs; physiotherapy; hydrotherapy, and complementary therapies. And putting your cat on a diet – however hard it might be for you both to stomach – can relieve symptoms and help to protect your cat's joints, which need all the help they can get, after all.

Hopefully, the case histories in this book will have demonstrated that a diagnosis of arthritis doesn't have to be the end of your cat's world – or yours. You'll still be able to do all the things you love doing together, and, by making some adjustments around the home to relieve the strain on your cat's joints – such as making sure your cat has a comfortable, warm bed; a litter tray that's easy to get in and out of, and food and water bowls at the right height – you'll be taking steps in the right direction.

Other ways to help manage the condition for the long term should include making sure your cat eats a healthy diet with all of the vitamins and minerals he needs, and visiting your vet if you see symptoms worsening, or you want to try another drug, perhaps with fewer side effects.

Although conventional treatment options might seem limited at the moment, and are largely centred on managing the pain of arthritis rather than offering a cure, take heart that research is under way into the role that genes play in arthritis. There's always the hope that gene-based therapy could be the treatment of the future – for people, dogs and cats at risk of developing the disease.

In the meantime, don't despair about an arthritis diagnosis: there's every reason to believe that your cat can still live her nine lives to the full!

Chapter 12

Useful contacts and further reading

Veterinary

The Cat Clinic
Melbourne, Australia
www.catdoctor.com.au

Willows Referral Service
Highlands Road
Solihull
West Midlands B90 4NH
England
Tel: 0121 712 7070
www.willows.uk.net

The Royal (Dick) School of Veterinary Studies
Edinburgh University
Easter Bush Campus
Midlothian EH25 9RG
Scotland
Tel: 0131 651 7300
www.ed.ac.uk/schools-departments/vet

Vet Professionals
www.vetprofessionals.com

Langford Veterinary Services (University of Bristol)
Langford House
Langford
Bristol BS40 5DU
England
Tel: 0117 928 9420; 01934 852 422
www.langfordvets.co.uk

Camboro Veterinary Hospital
Pennsylvania, USA
www.camboro.com

Dr Pierson
DrPierson@catinfo.org

Dick White Veterinary Referrals
Tel: 01638 572012
www.dickwhitereferrals.com

Physiotherapy

Sarah Long-Clemson
Tel: 07977 578156
www.vetphysio.com

Brian Sharp
caninephysio@yahoo.co.uk

Hydrotherapy

Avonvale Vet Centre
Wellesbourne

Warwickshire CV35 9NA
England
Tel: 01789 841072
www.avonvets.co.uk

Blossoms Animal Hydrotherapy & Physiotherapy
Blossoms Lane
Woodford
Cheshire SK7 1RE
England
Tel: 0161 439 3882
www.blossoms.co.uk

Hawksmoor Hydrotherapy Centre
Watling Street
Nuneaton
Warwickshire CV10 0TQ
England
Tel: 02476 350221
www.hawksmoorhydrotherapy.com

Meadow Farm Hydrotherapy & Rehabilitation Centre
Meadow Farm
North Common
Hepworth
Diss
Norfolk IP22 2PR
England
Tel: 01359 250310
www.meadowfarm.com

Canine Hydrotherapy Association (CHA)
www.canine-hydrotherapy.org

National Association of Registered Canine Hydrotherapists (NARCH)
www.narch.org.uk

Complementary

Osteopathy

Tony Nevin
Zoo Ost Ltd
29 Alstone Croft
Cheltenham
Glos GL51 8HB
England
Tel: 01242 221153
www.zooost.com

Nick Thompson
Holisticvet Ltd
Apthorp
Weston Road
Bath BA1 2VT
England
Tel: 01225 487778
www.holisticvet.co.uk

Acupuncture

Association of British Veterinary Acupuncturists (ABVA)
www.abva.co.uk

Trusts, clubs & charities

International Cat Care (formerly Feline Advisory Bureau) & International Society of Feline Medicine (ISFM)
High Street
Tisbury
Wiltshire SP3 6LD
England
www.icatcare.org
Tel: 01747 871872

Governing Council of the Cat Fancy
5 Kings Castle Business Park
The Drove
Bridgwater
Somerset TA6 4AG
England
Tel: 01278 427575
www.gccfcats.org
(The UK's main registration body for pedigree and non-pedigree cats)

The International Cat Association (TICA)
www.tica.org
(The worldwide feline genetic registry)

RSPCA
www.rspca.org.uk

North Norfolk Cats Lifeline Trust
www.northnorfolkcatslifelinetrust.co.uk

PDSA
Whitechapel Way
Priorslee
Telford
Shropshire TF2 9PQ
England
Tel: 01952 290999
www.pdsa.org.uk
(A veterinary charity caring for animals belonging to people in need)

Celia Hammond Trust (CHAT)
www.celiahammond.org
Tel: 01892 783367
(A charity running rescue and re-homing shelters for abandoned and abused cats)

Wood Green Animal Shelters
Kings Bush Farm
London Road
Godmanchester
Cambs PE29 2NH
England
www.woodgreen.org.uk
Tel: 0844 248 8181

Battersea Dogs & Cats Home
4 Battersea Park Road
London SW8 4AA
England
Tel: 0207 622 3626
www.battersea.org.uk

Blue Cross Pet Bereavement Support Service
Tel: 0800 096 6606
www.bluecross.org.uk

Winn Feline Foundation
www.stevedalepetworld.com
www.winnfelinehealth.org
(Contact: Steve Dale, certified animal behaviour consultant)

Maine Coon breed
www.lisieux.co.uk

Cleasanta Norwegian Forest Cats
www.cleasanta.co.uk

Bengal breed
www.typha-typhast.co.uk

Felis Britannica
www.felisbritannica.org
(Representing a federation of cat clubs across England, Scotland, Wales, Northern Ireland, the Channel Islands and the Isle of Man, all of whom wish to be part of the international association, Federation Internationale Feline (FIFe))

Online sites for prescription treatments, supplements, vet and pet products, and other services

www.snugglesafe.co.uk
www.vet-medic.com
www.animeddirect.co.uk
www.vetuk.co.uk
www.furrypharm.co.uk
www.petdrugsonline.co.uk
www.vmd.defra.gov.uk
www.vmd.gov.uk
www.medaust.com
www.petspec.co.uk
www.petelements.co.uk
www.healthforanimals.co.uk
www.petlifeonline.co.uk

Crematoria

Norfolk Pet Crematorium Ltd
Shortthorn Road
Felthorpe
Norfolk NR10 4DE
Tel: 01603 755438
www.norfolkpetcrematorium.com

Dignity Pet Crematorium
Odiham Road
Hook
Hampshire RG27 8BU
England
Tel: 01252 844572
www.dignitypetcrem.co.uk

Further reading

Magazines

Your Cat
BPG Media,
1-6 Buckminster Yard
Main Street
Buckminster
Grantham
Lincs NG33 5SB
England
Tel: 01476 859802
www.yourcat.co.uk

Forums & social networking sites

www.cataware.co.uk
www.mypetonline.co.uk
www.catchat.org
www.catsey.com

Annual cat events

Supreme Show of the Governing Council of the Cat Fancy
www.supremecatshow.org

The London Pet Show
www.londonpetshow.co.uk

All of the foregoing information was correct at time of going to press, and no responsibility is taken for omission or error. Inclusion does not confer endorsement by the author or publisher.

Rogues' Gallery

Rogues' Gallery

Index